AF564854

A Colour Handbook on Fundamentals of Entomology

NIPA® GENX ELECTRONIC RESOURCES & SOLUTIONS P. LTD.
New Delhi-110 034

About the Authors

Dr. Devinder Sharma, Double gold medalist (BHU Medal and Binani Gold medal) is presently working as Professor and Principal Investigator of AICRP (Honeybee and Pollinators), Faculty of Agriculture, Sher-e-Kashmir University of Agricultural Sciences-Jammu (J&K UT) India. He has specialization apiculture (bee management, disease and enemies, pollination management). He has authored 04 books, 08 manuals, 35 book chapters and published over 85 original research papers in various national and international journals. He has completed 10 externally funded projects on different aspects of beekeeping and is presently handling 04 externally funded projects. He is a recipient of the Best reporting centre award ICAR (2019)l Best Presenting Centre Award, ICAR (2022); Star Scientist Award (2022); Young Scientist award, Mahima Foundation (2021); Best Teacher Award, international Conference (2019), Kathmandu Nepal; First Prize for for his outstanding contributions in the field of Apiculture. He is also the recipient of certificate of merit from Jawahar Lal Nehru Foundation, New Delhi. He is the fellow of Entomological Society of India, editorial board member and Referee for scientific journals of national and international repute. He has successfully guided nine M.Sc students, two Ph, D students and currently guiding two (02) M.Sc and 02 Ph.D students.

Dr. Amit Kumar Singh is working as a professor, Division of Entomology, SKUAST-Jammu. He has significantly contributed in the field of IPM, acarology, horticultural and field crop protection. He has 18 years of research, education and extension experience in Agri. Entomology, and he has attended International and National seminars and conferences and he is a fellow and life member of many Professional Bodies/Societies. Dr. A.K. Singh has 45 research articles, 2 books, 15 book chapters, 19 conference papers, 2 technical reports, 12 manual chapters, 13 extension folders, and over 18 popular articles.

Dr. Rakesh Kumar Gupta is Professor and Head, Entomology and Professor and Head, Division of Sericulture, SKUAST-Jammu. He possesses good international exposure in term of biological control gained at Agricultural Research Organization, Israel under MASHAV Visiting Scientist Fellowship programme besides visiting twelve countries to update skills on pollination, IPM and agriculture through various scientific trips under VLIR UOS International Fellowship wherein he worked on biological control of greenhouse pests, Bee breeding and IPM. He also formulated the successful grant proposals and handled 11 Research Projects as Principal Investigator and 04 as Co- Principal Investigator as well. At the national level, He is acting as external examiner by prestigious institutes across the nation besides acting as an Expert Member, Commission for scientific and technical terminology: fundamental Glossary of Agriculture (English–Hindi-Dogri), Member, National Taskforce for sector programme of DBT Research,

Appraisal and funding for women specific projects. DBT (Competitive grants), Expert Member, Agricultural Sciences, India Science, Technology & Innovation, DST, Physical monitoring (on field) of national projects in DBT and expert Member-Project Review Committee-VARD (Agriculture) National innovation foundation, Ahmadabad. He also acted as National Convenor for designing ARS syllabus for Entomology, Plant Pathology and Nematology by ICAR and expert for CAS, ASRB and UPSC, New Delhi. He Has successfully guided Eleven (11) M.Sc students, Five (05) Ph, D students and currently guiding two (02) M.Sc and 03 Ph.D students. I possess good scientific skills on IPM, ecological methods, bio control methodologies, statistical methods and computational techniques. Above all I have the capacity to undertake interdisciplinary teamwork as well as to undergo independent research work.

Dr. Sachin S. Suroshe is working as a Project Coordinator of ICAR-AICRP (HB&P), Division of Entomology, ICAR-IARI, Pusa, New Delhi. Dr. Sachin S. Suroshe has significantly contributed in the field of Biological control, IPM, Apiculture, Pomegranate and Vegetable plant protection. Dr. Sachin has 20 years of research, education and extension experience in Agri. Entomology, and he has attended International and National seminars and conferences and he is a fellow and life member of many Professional Bodies/Societies. He has presented many guest lectures in national and international conferences. His field of interest is Biological Control, Insect Pest Management, and Apiculture. Dr. Sachin received many prestigious awards, viz., DST-SERB Young Scientist-2014; SARP (Society for the Advancement of Research on Pomegranate) Associate Award-2017; Scientist Award-2019 by Dr. B. Vasantharaj David Foundation, Chennai; Outstanding Scientist Award-2020 by New Age Mobilization Society, New Delhi; NESA eminent scientist of the year award-2021 by National Environmental Science Academy, New Delhi; Best Scientist Award-2022 by the Society of Plant Protection Sciences ICAR-NCIPM, New Delhi. Dr. Sachin has 57 research articles, 3 books, 5 book chapters, 15 conference papers, 5 technical reports, 23 manual chapters, 22 extension folders, and over 50 popular articles.

Kumaranag K.M. is presently working as Scientist (Senior Scale) in Project Coordinating Unit AICRP (Honeybee and Pollinators), IARI, New Delhi. He has specialization in IPM and apiculture. He is actively involved in developing methodologies for the experiments proposed in technical programme of AICRP (HB&P). He done his post-graduation from University of Agricultural Sciences, Bengaluru and Ph.D from CCS Haryana Agricultural University, Hissar. Dr. Nag has published 30 research articles, 3 book chapters, 10 conference papers, 3 technical reports, 8 manual chapters, 10 extension folders, and over 10 popular articles.

Kuldeep Srivastava is presently working as Principal Scientist (Entomology) at ICAR-IIVR, Varanasi having 20 years' experience in Teaching, Research and Extension activities. Dr. Srivastava did M. Sc. (Ag.) from CSAUA&T, Kanpur (1998) and Ph. D. from GBPUA&T, Pantnagar (2003) in Agricultural Entomology and started his career as Assistant Professor (Entomology) in 2004 at Sher-e-Kashmir University of Agricultural Sciences & Technology, Jammu, J&K and further joined as Senior Scientist (Ag. Entomology) at NRC on Litchi, Muzaffarpur, Bihar in 2013. After joining ICAR-IIVR, Varanasi in 2019 he is currently developing IPM modules for vegetable insect pests. He has handled 5 externally funded projects as principal investigator and guided 12 students as major advisor/ co- advisor. He is also credited with the development of several technologies/ varieties for insect pest management in horticultural crops. Dr. Srivastava is recipient of distinguished Scientist Award- Science & Tech society for Integrated Rural Improvement, Telangana; Fellow of Entomological Society of India (FESI), New Delhi; Fellow of Society of Plant Protection Sciences (FSPPS), New Delhi and Fellow of Confederation of Horticultural Association of India (FCHAI), New Delhi. and received various Awards/ Fellowship by reputed societies. He has published more than 70 research papers in reputed journals along with edited book (02), book chapters (25), monograph (01), training manual (03) and more than 70 popular articles.

A Colour Handbook on Fundamentals of Entomology

Devinder Sharma
Professor (Entomology)
Division of Entomology
Sher-e-Kashmir University of Agricultural Sciences & Technology of Jammu
Jammu, Jammu & Kashmir, India

Amit Kumar Singh
Professor (Entomology)
Division of Entomology
Sher-e-Kashmir University of Agricultural Sciences & Technology of Jammu
Jammu, Jammu & Kashmir, India

Rakesh Kumar Gupta
Professor and Head (Entomology)
Division of Entomology
Sher-e-Kashmir University of Agricultural Sciences & Technology of Jammu
Jammu, Jammu & Kashmir, India

Sachin S. Suroshe
Project Coordinator
AICRP (Honeybee and Pollinators)
IARI, New Delhi

Kumaranag K.M.
Senior Scientist
AICRP (Honeybee and Pollinators)
IARI, New Delhi

Kuldeep Srivastava
Principal Scientist (Entomology)
ICAR-IIVR, Varanasi, Uttar Pradesh

NIPA® GENX ELECTRONIC RESOURCES & SOLUTIONS P. LTD.
New Delhi-110 034

NIPA® GENX ELECTRONIC RESOURCES & SOLUTIONS P. LTD.

101,103, Vikas Surya Plaza, CU Block
L.S.C.Market, Pitam Pura, New Delhi-110 034
Ph : +91 11 27341616, 27341717, 27341718
E-mail:newindiapublishingagency@gmail.com
www: www.nipabooks.com

For customer assistance, please contact
Phone: + 91-11-27 34 17 17
Fax: + 91-11- 27 34 16 16
E-Mail: feedbacks@nipabooks.com

Print ISBN: 978-93-58871-24-1

ebook ISBN: 978-93-58871-25-8

Composed and Designed by NIPA®.

Preface

The book “A Colour Handbook on Fundamentals of Entomology" was conceived during the lockdown period of COVID-19. Although, there are a large number of standard and good books on the subject but they are either too exhaustive or too brief or mainly research oriented, and they deal, only one aspect of the subject, hence they do not fulfill the basic need of the under-graduate students of Entomology. Entomology is a biological science dealing with a specific group of organism, the insects. Insect constitute the largest class of the whole living organism. The general understanding of the subject of entomology is must for its practical application in the field. The book required by UG students strictly as per syllabus required by the under graduate students of Entomology is hardly available. The book embodies a fairly comprehensive treatment of the Basic Entomology. This book provides a balanced and up to date account on morphology, anatomy, physiology of various organs and systematics as well as systematics of class Insecta. A good number of illustrations and photographs have been introduced for better understanding of the students. The book would have useful to students preparation for various competitive examinations conducted by Indian Council of Agricultural Research-JRF, Agricultural Scientific Recruitment Board-ARS/NET, National Academy of Agricultural Research Management-SRF, Indian Agricultural Research Institute-PG/Ph.D. Entrance, Union Public Service Commission, SAUs Entrance and Allied Agricultural Examinations. We hope the Entomology students will find the book extremely useful and welcome this edition. We wish to thank all near and dears offering useful suggestions during the preparation of this study material.

Authors

Contents

1

Importance, Scope and History

Human beings came into existence 1 million years ago. Insects which constitute 70-90% of all animals present in this world came into existence 250- 500 million years ago.Insects are highly specialized group of invertebrates belonging to phyla arthropoda characterized by having a bilateral symmetrical body covered by Exoskeleton chitin; with jointed appendages provided with independently movable muscle and inserted in to special pockets of exoskeleton; with a haemocoelic body cavity containing the circulating blood in which are bathed the internal organs; with a dorsal heart and decentralized ventral nerve cord; and with tracheal system for respiration; metamorphosis direct or indirect.

The insects are bilaterally symmetrical tracheate Arthopods is which the body is divided into head, thorax and abdomen. A single pair of antennae is present and head also bears a pair of mandibles, a pair of maxillae,. The thorax possess three pairs of legs and usually two pairs of wings. The abdomen is devoid of any ambulatory appendages and the genital opening is situated near the posterior end of the body. Post embryonic development is rarely direct and metamorphosis usually occurs.

Insect definition

Insects (from Latin *insectum*, a calque of Greek ἔντομον [*éntomon*], "cut into sections") are a class of invertebrates within the arthropod phylum that have a chitinous exoskeleton, a three-part body (head, thorax and abdomen), three pairs of jointed legs, compound eyes and one pair of antennae.

What is Entomology

Entomology, from the Greek: *entomo-* "that which is cut in pieces or engraved/ segmented", hence "insect"; and *logos* "knowledge", is the scientific study of insects. At some 1.3 million described species, insects account for more than 2/3rds of all known organisms, dating back some 400 million years, and have many kinds of interactions with humans and other forms of life on earth, so it is an important specialty within biology. Though technically incorrect, the definition is sometimes widened to include the study of terrestrial animals in other arthropod groups or other phyla, such as arachnids, myriapods, earthworms, and slugs.

Like several of the other fields that are categorized within zoology, entomology is a taxon-based category; any form of scientific study in which the organisms studied happen to be insects is, by definition, entomology. Entomology therefore includes a cross-section of topics as diverse as molecular genetics, behavior, biomechanics, biochemistry, systematics, physiology, developmental biology, ecology, morphology, paleontology, anthropology, robotics, agriculture, nutrition, and more.

Entomology is a branch of zoology, which deals with the study of insect. The word entomology has two parts viz., Entomon = Insects and logos = study. Agril. Entomology deals with the study of insects in relation to agriculture

- Term Hexapoda was coined by Laterille (1825)
- Arthropod : Greek word
  - Entomology : Greek word
  - Insect : Latin word
- At some 1.3 million described species, insects account for more than 2/3rds of all known organisms, dating back some 400 million years,

Principles of Entomology

The field of entomology may be divided into 2 major aspects.

1. Fundamental Entomology or General Entomology
2. Applied Entomology or Economic Entomology

Fundamental Entomology deals with the basic or academic aspects of the Science of Entomology. It includes morphology, anatomy, physiology and taxonomy of the insects. In this case we study the subject for gaining knowledge on Entomology irrespective of whether it is useful or harmful.

Applied Entomology or Economic Entomology deals with the usefulness of the Science of Entomology for the benefit of mankind. Applied entomology covers the study of insects which are either beneficial or harmful to human beings. It deals with the ways in which beneficial insects like predators, parasitoids, pollinators or productive insects like honey bees, silkworm and lac insect can be best exploited for our welfare. Applied entomology also studies the methods in which harmful insects or pests can be managed without causing significant damage or loss to us.

In fundamental entomology insects are classified based on their structure into families and orders etc. in applied entomology insects can be classified based on their economic importance i.e. whether they are useful or harmful.

Economic classification of insects

Insects can be classified as follows based on their economic importance.

This classification us according to TVR Ayyar.

Insects of no economic importance:

There are many insects found in forests, and agricultural lands which neither cause harm nor benefit us. They are classified under this category. Human beings came into existence 1 million years ago. Insects which constitute 70-90% of all animals present in this world came into existence 250- 500 million years ago.

Insects of economic importance

A. Injurious insects

a) **Pests of cultivated plants (crop pests):** Each cultivated plant hatbours many insects pests which feed on them reduce the yield of the3 crop. Field crops and horticultural crops are attacked by many insect species. (eg) cotton bollworm, Rice stem bores.

b) **Storage pests:** Insects feed on stored products and cause economic loss. (eg) Rice wewil, Pulse beetle.

c) **Pest attacking cattle and domestic animals:** Cattle are affected by pests like Horse fly, Fleshfly, Fleas and Lice. They suck blood and sometimes eat the flash.

d) **House hold and disease carrying insects:** House hold pests include cockroach, ants, etc,. Disease carrying insects are mosquitoes, house flies, bed bugs, fleas etc.

B. Beneficial insects

(a) Productive insects

i) **Silk worm:-** The silk worm filament secreted from the salivary gland of the larva helps us in producing silk.

ii) **Honey bee:-** Provides us with honey and many other byproducts like bees wax and royal jelly.

iii) **Lac insects**:- The secretion from the body of these scale insects is called lac. Useful in making varnishes and polishes.

iv) **Insects useful as drugs, food, ornaments** etc,

(b) As medicine

- eg. Sting of honey bees- remedy for rheumatism and arthritis
- Cantharidin - extracted from blister beetle –useful as hair tonic.

(c) As food –

- for animals and human being.
- For animals- aquatic insects used as fish food.
- Grass hoppers, termites, Pupae of moths.

They have been used as food by human beings in different parts of the world.

(d) Ornaments, entertainers

- Artists and designers copy colour of butterflies.
- Beetles worn as necklace.
- Insect collection is an hobby

(e) Scientific research

- Drosophila and mosquitoes are useful in genetic and toxicological studies respectively.

Helpful insects

(i) **Parasites**

- These are small insects which feed and live on harmful insects by completing their life cycle in a host and kill the host insect.
- Eg egg, larval and pupal parasitoids

(ii) **Predators**

- These are large insects which capture and devour harmful insects.
- Eg Coccinellids, Preying matritids.

(iii) **Pollinators**

- Many cross pollinated plants depend on insects for pollination and fruit set.
- Eg Honey bees, aid in pollination of sunflower crop.

(iv) **Weed killers:**

- Insects which feed on weeds, kill them thereby killers.
- Eg Parthenium beetle eats on parthenium. Cochineal insect feeds in Opuntia dillenii.

(v) **Soil builders**

- Soil insects such as ants, beetles, larval of cutworms, crickets, collembola, make tunnels in Soil and facilitate aeration in soil. They become good manure after death and enrich soil.

(vi) **Scavengers**

- Insects which feed on dead and decaying matter are called scavengers.
- They important for maintaining hygine in the surroundings.
- Eg Carrion bettles, Rove beetles feed on dead animals and plants.

e) House hold and disease carrying insects

- Pests which cause damage to belongings of human being like furniture, wool, paper etc. Eg. Cockroaches, furniture beetle, Silverfish fish etc.
- Pests which cause painful bite, inject venoms. Eg. Wasps, bees sting us. Hairy caterpillar nettling hairs are poisonous. Mosquitoes, bugs bite, Pierce and suck blood from humans and pet animals.
- Disease causing Mosquito- Malaria, Filariasis, dengue fever. Housefly-Typhoid, Cholera, Leprosy, Anthrax

Importance of Entomology

Study of entomology aims at understanding insect body organization and function, their habitats, behaviours, relation to one another and to the surrounding in which they live, their classification, development distribution, post history and their economic importance.

Man has been attracted towards insects from very early times. The Sanskrit term Shatpada refers to the hexapodous condition of insects. Use of insects and insect products as items of medicinal value (oil beetles, use of lac, use of red ants in suturing wounds) have found a place in certain indigenous systems of treatments. Honey is used more as medicine than as a food. Parts of insects have been used as cheap jewellery by certain tribes.

As far as study of insect is concerned in agriculture it is of worth importance. Related to agriculture, insects are both useful and harmful. As a harmful role they damage our crops, lay waste vast areas of cultivated land, despoil on feed; they bring about different diseases and some of the examples are Malaria, plague, typhus, sleeping sickness etc. The serious disease malaria was earlier thought to be beyond the reach of mankind. A lot of money was just wasted in Malaria management but was of no use. When the entomologist studied the morphology and taxonomy of mosquitoes, they get interesting information on malaria transmitting mosquito, the *Anopheles* and the problem was solved just like anything. Another example is that of body louse spreading typhus fever in 1944 in Italy amongst the civilians living in crowded conditions. In 1939 Dr. Paul Muller discovered the insecticidal properties of DDT, which was then used in control of body louse. DDT become famous overnight as an insecticide and Muller saved thousands of soldiers from the deadly disease, typhus and not surprising, therefore that P. Muller was awarded the Nobel prize for Medicine in 1948. These are some insects, which are of international importance as a pest e.g. Locust. To show their devasting features some statistics related to them sufficient. A swarm of locust is estimated to contain more than 1500 million individuals. They fly with a speed of 150km/day. Even sunlight can not pass on earth when they are flying. They just eat everything which is green on earth. Army worms behaves like an army and destroy the crop within an overnight. This is about the harmful role played by insect. Now some of the useful roles played by the insects are.

- Insects are the cheap producers of some valuable natural commodities e.g. silk. Honey lac, cantharidine, tannic acid, formic acid etc.
- Insect act as natural pollinators.
- Insect plays viral role in biological control of other injuries insects and weeds.
- Insects improves the physical condition of soil by a) burrowing b) by adding their dead bodies.
- They play vital role in converting dead decaying organic material into organic manure.
- Some insects are valuable in scientific study e.g. Drosophila; insects helps in criminal investigations especially in murder cases (Housefly maggots)

Entomology as applied science has its own importance both in plant protection and getting benefit for mankind from the useful role played by insects. Scientific entomology is over a century old in India and a firm foundation of economic entomology has been laid by scientists like Lefroy, Fletcher, Ramkrishnan Ayyar and Ramchandra Rao in India. Till absent 1930, the subject was morphological all over the world. This study yielded considerable information on Taxonomy, insect pests and control. Towards the later period, applied entomology developed and a fundamental basis for pest study and methods of pest control including biological and legal control formed the major activity in this branch of science. Recognizing the usefulness of insect as an experimental animal Wigglesworth showed that physiology of insect is a fascinating field of study, and he established insect physiology as a separate discipline in 1930-38; in later period the same was extended into field of insect control also. In 1950-60 entomology emerged as a biological science encompassing morphology, ecology, physiology, phenology, pestology, biochemistry and ethology (behaviour).

Scope of Entomology

In their attempt to secure food insect inflict considerable damage to almost every part of the plant. Their capacity for multiplication and wonderful adaptations has made them a serious threat to human being and his existence. They compete man for three basic needs i.e. food, clothing and shelter. Among these food is common to both of them. Looking towards increasing population this vital item is now a days becoming scare and to secure this food man has to always think about insect as his first enemy. Due to this, entomology has its scope first in the area of plant protection. Research is continuously going on and newer insecticides are evolved as a result.

Still the problem is not solved. And hence man is taping new fields in plant protection. As result of this new generation pesticides were born some of them are use of hormones, antihormone compounds, chemosterilants, antifeedant, repellants, pheromones etc. which now a days replacing use of hazardous pesticides.

We are not yet able to identify most of the insects on earth where there may be chance of identifying a new pest. There is a scope for entomologist to identify more no of insects. Similarly scope is there for designing newer and cheaper methods of pest control because all the pesticides are toxic to human being causing several deaths every years. Also they are producing resistance in insect, where to control such a resistant progeny is a great challenge. To solve this problem upto certain extent a new technique adopted is IPM. (i.e. Integrated pest management) but still IPM for all the crops is not available here also a vast scope is there. New field called genetic engineering is coming up where one can develop pest resistant (immune) crop variety which will solve all further problems arising as a by product of pesticide use. Such work is in progress all over the world. (e.g Bt rice, transgenic cotton etc.)

On the other hand we haven't yet able to exploit the fullest utilization of useful role of insect in agriculture. As seen earlier insect benefit to us in several ways one of it is their valuable products. Still in India the silk industry is not growing as it is growing in other countries like Japan and France this is because of lack of knowledge, excluding Karnataka other states should come up in production of silk which has a great potential in earning foreign exchange. Similarly bee keeping it yields honey and wax which are very important commodities, other then that pollination by bees is an additional advantage and farmers should take help of these pollinators to augmenting and enhancing production and productivity of their crops their yields.

Entomology can generate employment up to certain extent e.g. production of parasites/ predators, diseases causing agents. This job one can do by taking short training and can earn money as well as can help the farmer in adoption cheap and non polluted plant protection. For increasing population we will be in short of food material and to secure it we have to do a lot of labour in this field . Where still a lot of scope is there.

Insect Biodiversity

- There may be over 50 million extant species on earth but only a small proportion of them have been described Robert May (1988), one of the distinguished evolutionary biologists. The scale of descriptive deficit varies widely, for smaller taxa such as insects with body size of 0.5 mm – 1.0 mm long that is far from complete.

- For instance there are some 1 million described species while the estimates vary from a very much conservative 2.75 million to 50 million.
- Largest order : Coleoptera
- Smallest order : Zoraptera
- As per records of Zoological Survey of India the entomo-fauna represent 6.04% of the global records.
- On an average 25.3% of the insect genera and 43.4% of species are endemic to the Indian region.
- Despite diverse habitats only about 46000 species of plant and 65,000 species of animals of which 55,000 are insects had been estimated to occur in India. This forms only a small percentage of known world biodiversity.
- Today, India occupying about 2% of global species documented nearly 7-8% of global faunal biodiversity
- Out of 86,874 animal species of the country, insects alone comprise 68.32%.

Organism	Total number of species	Per cent of total
Insects	1.1 million (Approx)	54
Higher plants	2.48 lakh	18
All other plants	1.07 lakh	7
All other animals	2.81 lakh	18
Others (Protozoa)	0.38 lakh	2

Insects are the longest occupants of our planet

Insect occupied more than 10% of the 3.5 billion years of earth's history of life

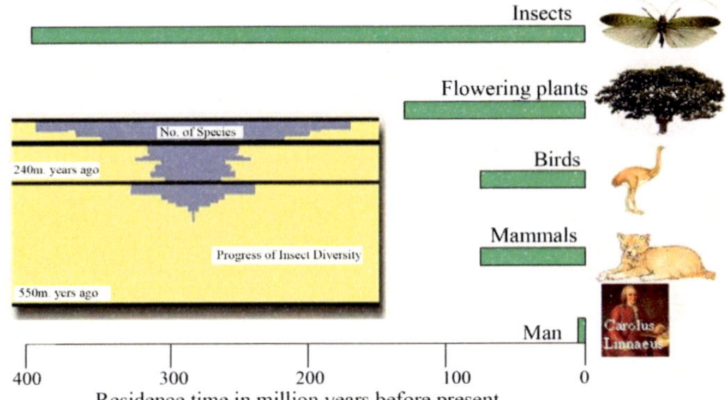

Insect Order	**Number of species**
Coleoptera (Beetles and weevils)	3,50,000
Lepidoptera (Butterflies and moths)	1,60,000
Hymenoptera (Bees, wasps and ants)	1,20,000
Diptera (Flies and mosquitoes)	1,20,000
Hemiptera (Bugs)	98,000
Orthoptera (Grasshopperas, crickets and locust)	20,000

The five kingdom classification of organisms
(according to Margulis and Schwartz)

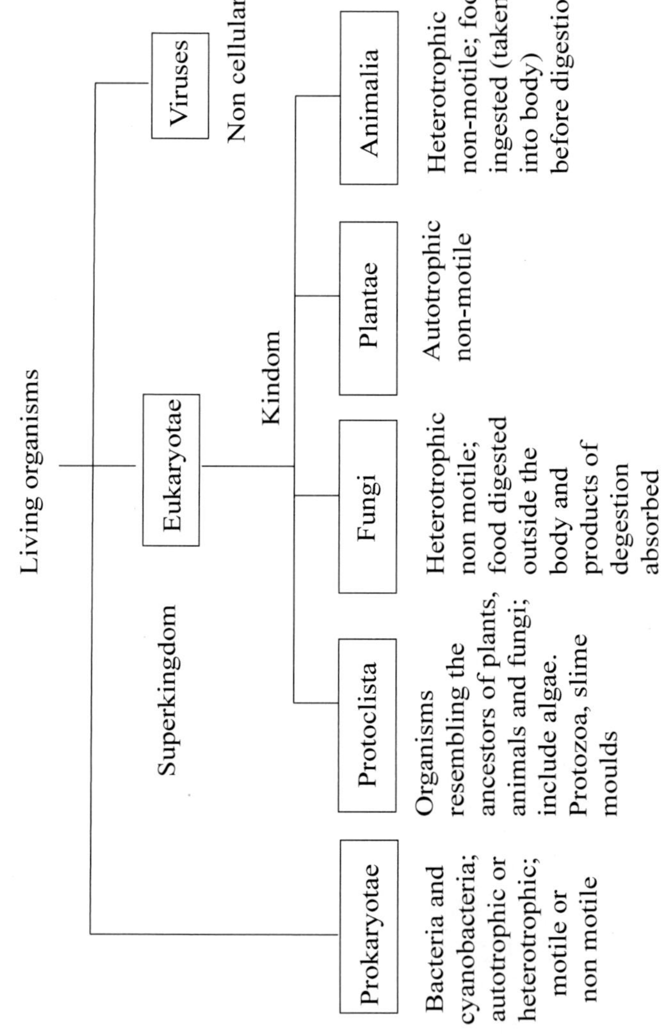

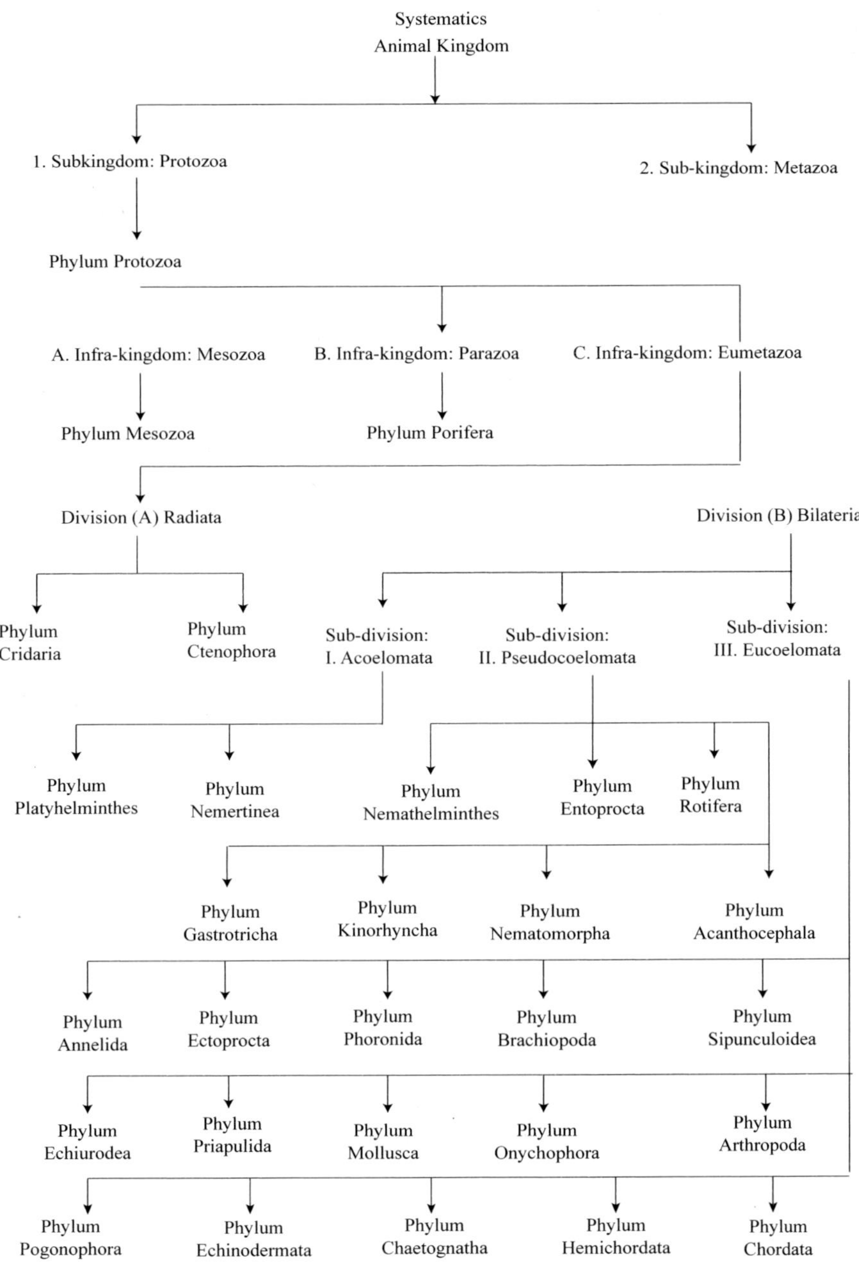
Systematics
Animal Kingdom
1. Subkingdom: Protozoa
2. Sub-kingdom: Metazoa
Phylum Protozoa
A. Infra-kingdom: Mesozoa
B. Infra-kingdom: Parazoa
C. Infra-kingdom: Eumetazoa
Phylum Mesozoa
Phylum Porifera
Division (A) Radiata
Division (B) Bilateria
Phylum Cridaria
Phylum Ctenophora
Sub-division: I. Acoelomata
Sub-division: II. Pseudocoelomata
Sub-division: III. Eucoelomata
Phylum Platyhelminthes
Phylum Nemertinea
Phylum Nemathelminthes
Phylum Entoprocta
Phylum Rotifera
Phylum Gastrotricha
Phylum Kinorhyncha
Phylum Nematomorpha
Phylum Acanthocephala
Phylum Annelida
Phylum Ectoprocta
Phylum Phoronida
Phylum Brachiopoda
Phylum Sipunculoidea
Phylum Echiurodea
Phylum Priapulida
Phylum Mollusca
Phylum Onychophora
Phylum Arthropoda
Phylum Pogonophora
Phylum Echinodermata
Phylum Chaetognatha
Phylum Hemichordata
Phylum Chordata

History of Entomology

- The knowledge about insects dates back to 6000 yrs as Indian ancestors were well versed in the art of rearing silk worms and weaving silk cloth. Even during 3870 BC an Indian king sent various silken materials as presents to a Persian king. Reference to the lac (laksha) and dye found in ancient Sanskrit works like Chanakya sutra, Arthasastra and Sakuntala, show that the people were aware of the usefulness of lac. The mythological epics, the Ramayana (2550-2150 BC) and Mahabharata (1424-1366 BC) mentions about silk, honey and lac. The first detailed classification of insects was done by **Umaswati** (0-100 AD). Classification of bees by the Indian physician **Charaka** (1200-1000BC) and classification of ants, flies and mosquitoes by the surgeon **Sushruta** (100-200 AD) are the evidences for our earliest knowledge about insects. Susrutha even classified ants (pipilika) in to six groups and gave details of the structure and habits of these insects.
- The reference to insects were found in Sanskrit dictionary, 'Amarakosa' and in books like Artha Sastra, Chanakya Sutra etc. In ancient scripts like Ramayana and Mahabharata, some of the terms used were related to insects. They are:

  Pipilika – Ant,

  Pathanga – grasshoppers,

  Madhumakshika – honey bees,

  Umbakapalika - termite queen

Aristotle taught entomology to his pupils in Athens. He made investigation about flies, honey bees, ants etc.

The first true morphologist apparently was **RÈnÈ Antoine Ferchauld, Seigneur de RÈaumur, des Alpes et de la BermondiÈre** (1683-1756). They published six volumes on the life histories of a great diversity of insects.

Karl de Geer (1720-1778), a Swedish, illustrated with great exactness many structures and organisms for the first time, including Collembola with descriptions of the collophore and other structures.

Charles Bonnet (1720-1793), provided descriptions of caterpiller respiration and aphid parthenogenesis. He coined the term "insectology" to refer to studies of all animals that were not molluscs or vertebrates

Pieter Lyonet (1707-1789), a Dutch lawyer studied the anatomy of the moth, *Cossus ligniperda*, publishing 600 pages and 18 engraved plates on the subject in 1762.

Aldrovandi built a system of classification on the basis of leg and wing morphology in 1602,

Mouffet used the same characters for a different system in 1634, and

Swammerdam described kinds of metamorphosis in 1669.

With his publication of *Systema Naturae* in 1758, **Linnaeus** established the starting date for binomial nomenclature, and used wing morphology to organize insects into his classification scheme. This contained the earliest record of 28 species of Indian insects.

Fabricius (1745-1808) Danish scientist devised a much more influential classification in many volumes, including generic and species descriptions, and separate volumes on each order that he recognized at the time. His work was concentrated in the field of descriptive systematics. His enormous importance to comparative morphology and systematics lies in his descriptions, his grouping into genera, and his systematizing based on mouthparts. He classified insects into 3 orders based on the type of mouth parts. The first entomologist who made an extensive study of Indian Insects.

Studies by **Lamark** (1809), **CuviÈr** (1816), **Leuckart** (1848), **Vogt** (1851), **Haeckel** (1874), **Hatschek** (1888), and many others led to what we now recognize as the major phyla of metazoan life and contributed to the placement of arthropods as one of the two major phyla of protostome metazoans, the other being the Mollusca. As these ideas took form, arthropod morphology became intimately linked to systematics and taxonomy.

Comstocks, **John Henry** (1849-1931) and **Anna Botsford** (1854-1930) wrote several influential textbooks and worked on a variety of insects, contributing information on morphology and anatomy for most of them. Their most lasting contribution to insect morphology and anatomy was the study of wing veination amony orders and their efforts to standardize wing vein terminology. With the exception of the Hymenoptera, a modification of their system, called the Comstock-Needham system, is still widely used today.

The best known insect anatomist and morphologist of the 20th Century was **Robert Snodgrass** (1875-1962). His most celebrated contributions to insect morphology and anatomy are: *Anatomy and Physiology of the Honeybee* (1925), *The Principles of Insect Morphology* (1935), and *Textbook of Arthropod Morphology* (1952).

Chapman and **Matsuda**. **R. F. Chapman's** *The Insects, Structure and Function* was first published in 1969, with the latest,5th E edition serving as basic textbook in morphology.

Ryuichi Matsuda worked first out of Theodore Hubbel's lab at the University of Michigan, then at the Biosystematics Research Institute in Canada. He published three monumental volumes dealing the the comparative anatomy of the head, thorax, and abdomen. Matsuda took a modern, analytical phylogenetic approach to comparative relationships even before the adoption of Hennig's cladistic methods.

H. Bruce Boudreaux, a mite specialist, was an outstanding phylogenetically oriented morphologist and published *Arthropod Phylogeny with Special Reference to Insects* around 1980, making it one of the first treatments of a major phylum from a phylogenetic approach.

First edition (1939) of **V.B. Wigglesworth's** *Principles of Insect Physiology* was published.

History of Entomology in India

- Hymns of *Atharva Veda* on the control of insects and crops reflect the variety of pests. *Manudharma sastra* (1000 BC) identifies bees, biting insects like mosquitoes and ants,
- The treatises of Charaka (1200 BC) on bees and *Shushmita's* (100–200 AD) work on insect stings and his classification of ants (*Piplika*), flies (*Mahashikala*) and mosquitoes (*Mashakala*) are of interest.
- Marasimha nearly 1000 years before Latrielle, coined the term 'shatpada' for hexapoda or six-legged animals. More detailed classifications by Umaswati and usage of the terms *ayonija* (ovipary) and *yonija* (vivipary) and *amba kapalika* for the termite queen were followed by the adoption of terms like *Indragopa* for cochineal, as well as the distinction between locusts (*salabha*), grasshoppers (*Patanga*) and thrips (*Trinapatra*).
- The first entomologist who made an extensive study of Indian insects was J. C. Fabricius (1745–1808) and the publication of Carl Linnaeus' (1758), *Systema Naturae* (10th edition) confirmed the earliest record of Indian insects, with 28 species.
- The immaculate drawings of Edmund Donovan (1800) reflect his power of observation. published an illustrated book entitled "An epitome of the natural history of insects of India and the Islands in the Indian seas" which was the first pictorial documentation on the insects of Asia and was revised in 1842 by West Wood.
- Linnaeus' student J.G.Koenig (1728–1785) who accomplished scientific work in India on insects. collected number of insects from Coromandel area and Southern Peninsular India and his collections were studied and named by Professor Linnaeus himself. He also published a special account of the termites of Thanjavur District. Fabricius, made Koenig's name remembered for ever by naming the well known and destructive red cotton bug of this country as *Dysdercus koenigi*.
- Westwood (1847) *Cabinet of Oriental Entomology* comprises a selection

of some of the rarer and more beautiful species of insects native to India, now housed in the The Hope Entomological Collections, Oxford University Museum of Natural History, Oxford.

- James Frederick William Hope published the *Entomology of Himalaya and India*, their entomo geographical characters emphasizing that Himalayan genera closely approximate Siberian forms and that entomological characters of a country are particularly influenced by temperature, vegetation and soil.
- 1782: "Dr. Kerr" published an account of Lac insects. In 1790, a detailed account of lac insect was published in the Asiatic Researches by the botanist Roxburgh.
- 1785- Asiatic Society of Bengal started in Calcutta by Sir William Jones.
- 1791- Dr. J. Anderson issued a monograph on Cochineal scale insects.
- 1799- Dr.Horsfield, an American doctor and first Keeper of the East India Museum published his famous book "A catalogue of the Lepidopterous Insects in the Museum of the Honourable East India Company, 2 vols. (1857, 1858-59).
- 1800- Buchanan (Traveler studying the natural wealth of India) wrote on the cultivation of lac in India and on sericulture in some parts of South India.
- 1897: Sir Ronald Ross , an IMS officer, in charge of a Madras regiment stationed in Begumpet (Secunderabad) made the discovery of the malarial parasite in a dissected *Anopheles* mosquito On August 20, 1897. August 20 is World Mosquito Day.
- 1875- Foundation of the "Indian Museum" at Calcutta. The "Indian Museum Notes" in five volumes (1889-1903), which contributed much on economic entomology and applied entomology in India. 19th Century marks the major progress and expansions in the field of applied entomology.
- 1883- "Bombay Natural History Society" was started. After the foundation of these two organisations scientific studies received greater attention in India. Numerous contributions of Indian insects were published in the Journal of the Bombay Natural History.
- 1883- Commencement of "Fauna of British India" series under the editorship of W.T.Blandford.
- 1892- Entomological part of the "Fauna of British India" (now Fauna of India) series started with Sir George Hampson contributed first of the four volumes on the moths of India.

- Dr. Rothney" (1848–1922) published the book 'Indian ants' (1893) (Earliest record of biological pest control in India ie white ants attack on stationery items kept free by red ants. His greatest collection of Oriental Hymenoptera also housed in the Hope Department of Entomology, as well those of Swinhoe (1836–1923) on Indian Lepidoptera are considered entomological treasures.
- 1901 : "Lionel de Niceville" was appointed as first entomologist to Govt. of India.
- Forest entomologist, Percy Stebbing (1870– 1960) whose pioneering work on Indian forest insects continues to be an important source of a work of reference. He was the first imperial forest entomologist published "Indian forest insects of economic importance :coleopteran
- Maxwell Lefroy, the first Imperial entomologist whose monumental volume on '*Indian insect pests*' (1906) and *Indian Insect Life* (1909) replete with information on economic aspects of entomology, became the only source book of information for well over 2 decades.
- Bainbrigge Fletcher (1878– 1951) the first Govt Entomologist of Madras state (1912), publish *Some South Indian Insects* and *Veterinary Entomology for India*, besides the *Catalogue of Indian Insects* which ran to 25 parts.
- Punjab and Uttar Pradesh government appointed their first State/ Provincial Government Entomologists in 1919 and 1922, respectively
- T. V. Ramakrishna Ayyar (1880–1952) : *Handbook of Economic Entomology for South India* is still much sought after, and he is rightly termed as the father of insect taxonomy in India (1940).
- Y. Ramachandra Rao (1885–1972) an acknowledged expert on locusts, their ecology and behaviour and his comprehensive monographs on locusts "*The Desert Locust in India*" (1960) are rare works of excellence, hailed by the international community, in particular by the International Locust Research Centre. Locust Warning Organisation (1939) was established in India. He was instrumental in founding Entomological Society of India, served as Founding Editor for Indian Journal of Entomology. He was given Rao Bahadur and Diwan Bahadur by British Government.
- Other famous entomologists as the late Hem Singh Pruthi, S. Pradhan, N. C. Pant, M. L. Roonwal, A. P. Kapur, E. S. Narayanan and others whose contributions in diverse fields have raised the quality of entomological research in this country during the post-independence period. Special mention has to be made of the doyen of Indian entomology,

- M. S. Mani's Entomological ventures in the Himalayas led him to publish his treatise on *High Altitude Entomology*. Being an acknowledged expert in the field of ceccidology, his classic works on *Ecology of Plant Galls* is a work of reference. He also published "*General Entomology*".
- The role of the entomological meetings of Pusa (1915–1923) has been particularly emphasized, since the proceedings of the meetings have been a source of abundant information on many aspects of insect pests, beneficial parasites, and predators, insects of industrial importance, medical and vector role of insects and will ever be a source of inspiration to entomologists.

S. Pradhan

- Carried out studies on economic entomology and insecticide toxicology that led to more of basic research in understanding mode of action of insecticides and fate of insecticides in the environment.
- He was the first Professor of Division of Entomology, IARI
- Pradhan's contribution to identification of the insecticidal principles of neem,
- the concept of Integrated Pest Management (IPM),
- periodicity of locust swarms,
- mode of action of DDT and
- Development of grain storage structure **'Pusa Bin'** to prevent losses due to stored insect pests stand out prominently in the annals of entomological research of our country.
- Published book "Insect Pests of Crops" (1969) Biometer and Biotic theory on prediction of locust outbreaks can be included

Hem Singh Pruthi

- 1st Indian Imperial Entomologist of India (1934) succeeded Fletcher
- 1st Chief Editor of Indian Journal of Entomology (1939) published by Entomological Society of India (1938). Afzal Hussain was the first President of the Entomological society of India and Hem Singh Pruthi and Ramakrishna Ayyar were the the Vice-presidents of the society.
- 2nd President of Entomological society of India
- 1st Plant Protection Advisor to Government of India (1946)
- 1st Director of Directorate of Plant Protection Quarantine and Storage (DPPQ&S, Faridbad) in 1946
- Published book "Text book of Agricultural Entomology" (**1963**)

N.C. Pant

- Studied nutritional requirements of insects and the role of symbionts.

E.S. Narayanan and B.R. Subba Rao

- Established a strong unit of biological control, with an emphasis on taxonomy of parasites and predators.

Mehrotra, K. N.

- Provided a biochemical basis especially differential acetyl cholinesterase inhibition and carboxylesterase activity for insecticide selectivity and resistance, respectively.

T.N. Ananthakrishnan **(1969)** The monograph on Indian Thysanoptera

B. Vasantharaj David and **T. Kumara Swami** (1975): Book *Elements of Economic Entomology*

M R G K Nair (1975) book : *Insects and Mites of crops in India*

K.K. Nayar, N. Ananthakrishnan and **B. Vasantharaj David** (1976): Book *'General and applied Entomology'*

- **1912**- Plant Quarantine Act was enforced.
- **1914**- Destructive Insects and Pests Act was enforced.
- 1937 : A laboratory for storage pests was started at Hapur, U.P.
- 1937: Entomology division was started in IARI, New Delhi
- 1939-establishment of locust warning organization after the locust plague during 1926-32
- **1946**- 'Directorate of Plant Protection, Quarantine and Storage' of GOI started.
- 1953: National Malaria eradication programme was launched
- **1960**- "The Desert Locust in India" monograph by Y.R. Rao.
- **1968**- The Govt. of India enacted 'Central Insecticide Act'which came into force from 1st January, 1971.
- **1992** : The Destructive Insect and Pests (Amendment and Validation) Act
- **2003** : Plant Quarantine Order

At present, the Zoological Survey of India (ZSI), Kolkata, Indian Agricultural Research Institute (IARI), New Delhi, Indian Institute of Science (IISC), Bengaluru , Forest Research Institute (FRI) of Dehradun

Establishment of Entomological Institutes

S.no	Year	Institute	Location
1	1905	Establishment of "Agricultural research institute" by Lord Curzon. The land was donated by Mr.Phiffs of USA after whom the place was named as Pusa. Professor Lefroy became the first Imperial Entomologist. He convened a series of entomological meetings on all India basis to bring together all the entomologists of the country. From 1915 five such meetings were held at the Imperial Agricultural Research Institute at Pusa. While the Proceedings of the first meeting was not published, proceedings of subsequent four meetings became a treasure of entomolgical knowledge, which can never be overlooked by any student of Indian Entomology.	Pusa, Bihar
2	1906	Forest research institute	Dehradun
3	1911	"Agricultural research institute" Pusa renamed as Imperial Agricultural Research Institute	Pusa, Bihar
4	1914	Zoological survey of India	Kolkata
5	1925	Indian Institute of Natural Resins and Gums (**IINRG**) formerly known as Indian Lac Research Research Institute (**ILRI**) Now it is National Institute of Secondary Agriculture	Namkum, Ranchi
6	1936	Imperial Agricultural Research Institute, Pusa shifted to New Delhi	New Delhi
7	1937	Establishment of Entomology Division at IARI	New Delhi
8	1946	Directorate of Plant Protection Quarantine and Storage(DPPQS)	Faridabad
9	1946	National Institute of Plant Health Management, **NIPHM** (formerly known as Central Plant Protection Training Institute (**CPPTI**), later National Plant Protection Training Institute (**NPPTI**)	Hyderabad
10	1947	Imperial Agricultural Research Institute,renamed as Indian Agricultural Research Institute	New Delhi
11	1993 as PDBC	National Bureau of Agricultural Insect Resources (**NBAIR**) formerly (National Bureau of Agricultural Important Insects) (**NBAII**) OR Project Directorate on Biological Control (**PDBC**)	Bangalore
12	1988	National Institute of Integrated Pest Management (Earlier NCIPM)	New Delhi

Arthropod Evolution

Geological timetable

Million Year ago	Geological period	Events that occurred for plants and animals	
		Plant life	**Animal life**
0	Cenozoic era-modern life		
1	Quarternary	Flowering plants	Modern insect life
63	Tertiary	Angiosperms	Lepidoptera diversified
135	Mesozoic era Cretaceous	Flowering plants	
181	Jurassic	Gymnosperms	Isoptera, Lepidoptera, diptera, social insects
230	Triassic		Hymenoptera
280	Paleozoic era Permain	Gymnosperms	Thysanoptera, Hemiptera, Coleoptera
350	Carboniferous	Mosses	Orthoptera,
405	Devonian	Algae , ferns	Beginning of land animals
425	Silurian	Rise of land plants	Marine invertebrates
500	Ordovician	Algae	First chordates
600	Cambrian	Algae	Invertebrates

The Monophyletic Hypothesis

All arthropods differ from their annelid ancestors in several important ways: they have a chitinous exoskeleton, jointed/segmented appendages, a well-developed head and mouthparts, striated muscles, and an open circulatory system with a dorsal heart. Unlike annelids, the arthropods produce large, yolk-laden eggs that are encased in a proteinaceous shell. Ciliated nephridia, paired segmental excretory organs found in annelids and onychophorans, have been replaced in arthropods by specialized excretory organs located on the head (green glands in crustacea), near the legs (coxal glands in horseshoe crabs), or in the abdomen (Malpighian tubules in terrestrial arthropods). Remarkable similarities in the chemical composition of the exoskeleton and in the ultrastructure of the compound eyes in organisms as diverse as millipedes, shrimp, and horseshoe crabs provide strong evidence that all of these groups (myriapods, crustaceans, and chelicerates) must have evolved from a common ancestor.

- **Boudreaux** Theory : recognizes three evolutionary lineages within the phylum Arthropoda: Trilobita (extinct), Chelicerata, and Mandibulata
- Chelicerata includes all arthropods with fang-like mouthparts (chelicerae): spiders, ticks, and mites (Arachnida), horseshoe crabs (Xiphosura), and sea spiders (Pycnogonida). Mandibulata includes all arthropods that have chewing mouthparts (mandibles): crustacea, myriapods, and insects.
- **Handerlisch (1908,17)** : Triboliata theory
- **Sharov (1966)** : Arthropods arise from Annelids

- **Imm, A.D.** : Symphyla theory
- **Hansen** : Insects originate from crustacean group Syncarida
- **Crampton :** Insects originate from Isopods
- **Meglitsch and Schram (1991)** regard the Onychophora as closely related to myriapods. In their monophyletic classification scheme, Crustacea and Chelicerata are the most primitive groups. Insects, myriapods, and onychophora are grouped together in the superclass Uniramia because of similarities in leg structure and locomotion.
- **Jarmila Kukalová-Peck**, a Canadian entomologist, has proposed that "primitive" arthropods may have had as many as eleven segments in each walking leg.

The Polyphyletic Hypothesis

- **Charlotte Manton,** one of the founders of the polyphyletic hypothesis, suggested that there is a fundamental difference between the appendages of crustacea and those of other arthropods such as insects and myriapods (millipedes and centipedes).
- Manton argued that crustacean appendages are biramous; that is, two apical units (rami) are attached to a single basal unit. Appendages of other arthropods are uniramous: a single apical segment is attached to a single basal segment. Manton believed that crustaceans evolved from annelid worms similar to marine polychaetes of today, and that all other arthropods evolved from annelid worms that were more similar to the onychophora.
- **D. T. Anderson** whose studies of arthropod eggs has revealed that initial cell division in crustacean embryos is holoblastic (spiral cleavage), whereas the eggs of all other arthropods are meroblastic (superficial cleavage). The eggs of all known annelids are holoblastic.

Tools and Techniques

- **Fossil Record** Paleontology is the study of the history of life on Earth as based on fossils. Fossils are the remains of plants, animals, fungi, bacteria, and single-celled living things that have been replaced by rock material or impressions of organisms preserved in rock. Fossils can also provide evidence of the **evolutionary** history of organisms. **Paleontologists** use fossil remains to understand different **aspects** of **extinct** and living organisms. Individual fossils may contain information about an organism's life and **environment**. Insects have a relatively extensive fossil record, with approximately 1,263 families in 30 commonly recognized orders having been identified. Although only a few fossil insects (such as Collembola) are known from the lower Devonian, a massive radiation began sometime during the early Carboniferous, period more than 325 million years ago (mya). The first insects lack wings, but winged insects (the pterygotes) soon radiated into stem groups of all major lineages, including ephemeroids, odonatoids, plecopteroids, orthopteroids, blattoids, hemipteroids, and endopterygotes. The number of orders present by the upper Permian period was similar to that of today although some of the orders are now became extinct and new orders have evolved.

- **Radiometric dating** Carbon is a very special element. In combination with hydrogen it forms a component of all organic compounds and is therefore fundamental to life. Willard F. Libby of the University of Chicago predicted the existence of carbon-14 before it was actually detected and formulated a hypothesis that radiocarbon might exist in living matter. Willard Libby and his colleague Ernest Anderson showed that methane collected from sewage works had measurable radiocarbon activity whereas methane produced from petroleum did not. Perseverance over three years of secret research to develop the radiocarbon method came into fruition and in 1960 Libby received the Nobel Prize for chemistry for turning his vision into an invaluable tool. Carbon has three naturally occurring isotopes, with atoms of the same atomic number but different atomic weights. They are ^{12}C, ^{13}C and ^{14}C. C being the symbol for carbon and the isotopes having atomic weights 12, 13 and 14. The three isotopes don't occur equally either, 98.89% of carbon is ^{12}C, 1.11% is ^{13}C - and only one ^{14}C atom exists in nature for every 1,000,000,000,000 ^{12}C atoms in living material. ^{12}C and ^{13}C are both stable whereas ^{14}C is unstable or radioactive. Libby and his colleagues first discovered that this decay occurs at a constant rate. They found that after 5568 years, half the ^{14}C in the original sample will have decayed and after another 5568 years, half of that remaining material will have decayed, and so on. This half-life (t 1/2) is the name given to this value which Libby measured at 5568,30 years. This became known as the Libby half-life. After 10 half-lives, there is a very small amount of radioactive carbon present in a sample. At about 50 000 to 60 000 years, the limit of the technique is reached (beyond this time, other radiometric techniques must be used for dating). By measuring the ^{14}C concentration or residual radioactivity of a sample whose age is not known, it is possible to obtain the number of decay events per gram of Carbon. By comparing this with modern levels of activity (1890 wood corrected for decay to 1950 AD) and using the measured half-life it becomes possible to calculate a date for the death of the sample

- **Numerical Taxonomy** The polythetic concepts contrast with traditional monothetic analysis (Mayr 1969). All phenetic characters are given equal weight (Véron 1969), and are not given *a priori* discriminant value.

Simultaneous consideration of all characters gives each phenon its identity and hierarchical position. Using automatic data processing methods, phenetic analysis shows relationships between groups as graphs of factor analysis, or as dendrograms.

Cladistic analysis indicates apomorphic characters and determines the most parsimonious evolutionary sequence. The final product of the analysis is a cladogram which enables hypotheses of phylogenetic relationships to be made and indicates the direction of evolutionary change in characters.

- Statisticaltechniquesemployboth"primitive"and"advanced"Characteristics (the phenetic approach) or rely exclusively on "advanced" characteristics (the cladistic approach). In either case, the resulting "phylogenetic trees", variously known as phenograms, cladograms, or dendrograms, are intended as visual representations of perceived biological affinity.
- **Biochemistry** The first insect genome assembly (*Drosophila melanogaster*) was published two decades ago. Today, nuclear genome assemblies are available for a staggering 601 insect species representing 20 orders. Although representing just 0.06% of the ~1 million described insects, this breadth of insect genome sequencing still spans ~480 Myr of evolution and roughly two orders of genome size from the tiny 99 Mb genome of *Belgica antarctica* to the massive genome of *Locusta migratoria* at 6.5 Gb.
- Enzymes and metabolic pathways can reveal inherent patterns in nutrient processing, chemical defense, locomotion, intercellular and intracellular communication, or homeostatic mechanisms. Similarities and/or differences in these basic life processes are often consistent within an evolutionary lineage. The sudden appearance of a novel biochemical mutation may confer selective advantages and eventually lead to adaptive radiation of new species from a common ancestor.
- **Nucleotide Sequencing**

Insect Dominance

The dominance of insects is true in terms of diversity (number of species), numbers of individuals, and total biomass within a given area. The known number of insect species is only a fraction of the estimated total number of species. The total number of insect species has been estimated to be between five and ten million species; most of which have yet to be discovered and scientifically described. When all of the described species on the planet are considered, the number of insect species accounts for more than 50% of the total (Fig. 1). As more species are discovered, the proportion of insects is likely to increase as our knowledge of the biodiversity of other plant and animal species is much more complete. The overwhelming majority of insect species are not harmful to human health, commerce, or agriculture, and in some form or fashion are actually beneficial. World wide, less than 1% of all known insect species are pests either destroying/damaging food and fibre resources, stored products, structures, or transmitting diseases. Within forest ecosystems, several insect species are serious pests that threaten our use of these natural resources and the ecosystem services they provide. In short, insects are our greatest competitors for the resources required to sustain our lives. Those few species that are pests result in tremendous efforts to manage their populations and limit the damage they cause.

Measures of dominance

1. More number of species
2. Large number of individuals in a single species: e.g. Locust swarm comprising of 109 number of individuals, occupying large area.
3. Great variety of habitats
4. Long geological history

- Insects are the most dominant species on the earth as they originated on earth 480 million years ago. Among 1.7 million living species, 0.95 million species are insects.

Insect Order	Number of species
Coleoptera (Beetles and weevils)	3,50,000
Lepidoptera (Butterflies and moths)	1,60,000
Hymenoptera (Bees, wasps and ants)	1,20,000
Diptera (Flies and mosquitoes)	1,20,000
Hemiptera (Bugs)	98,000
Orthoptera (Grasshoppers, crickets and locust)	20,000

Reasons for Dominance

Insects constitute the largest class not only of the animal kingdom but of the living world as a whole including both the animal and the plant kingdom. Insects account for more than two-thirds (about 70%) of all known organisms date back some 400 million years (estimated between 250-500 million years ago), much earlier than appearance of humans, who only appeared 1 million years ago. There are about 1.3 million (=13 lakhs) described species. Insects constitute the most dominant class of the animal kingdom. They also represent the culmination of evolutionary development in terrestrial arthropod.

Exoskeleton

- The insect body has an outer exoskeleton or body wall made up of cuticular protein called as Chitin is polysacchride. The exoskeleton not only gives shape and support to the body's soft tissues, but also provides protection from attack or injury, minimizes the loss of body fluids in both arid and freshwater environments, and assures mechanical advantage to muscles for strength and agility in movement.

- The exoskeleton can resist both physical and chemical attack. It is covered by an impervious layer of wax that prevents desiccation.
- Freedom of movement is ensured by membranes and joints in the exoskeleton. Muscles that attach directly to the body wall combine maximum strength with near-optimum mechanical advantage (leverage). The result is an ant, for example, that can lift weight up to 50 times its own body weight.

Small Size

- Most species are between 2 and 20 mm (0.1 - 1.0 inch) in length, although they range in size from giant moths that would nearly cover your computer screen to tiny parasitic wasps that could hide inside the point at the end of this sentence.
- *Dicopomorpha echmepterygis* is the smallest of the small. Discovered in 1997, this Costa Rican wasp (family Mymaridae) is a parasite of other insects' eggs. Adult males may be only 0.139 mm (0.00055 inch) in length -- nearly 1/3 smaller than some single-celled protozoa (e.g., *Paramecium caudatum*).
- Small size is a distinct advantage.
    - More number of individuals per unit area
    - Their movements require a smaller muscle volume and consume less energy.
    - Minimal resources needed for survival and reproduction.
    - Small size is a big advantage to insects that must avoid predation.

Flight

- Flight gave these insects a highly effective mode of escape from predators that roamed the prehistoric landscape.
- It was also an efficient means of transportation, allowing populations to expand more quickly into new habitats and exploit new resources. It also helps to
    - To colonize in new sources of food
    - To locate shelter
    - To find their mate
    - To escape from enemies

    - To locate suitable egg laying site
    - To escape from unfavourable situations

- Although the metabolic cost of flight (calories per unit of lift) is similar to that of birds and bats, an insect's flight musculature produces at least 2X more power per unit of muscle mass (see chart below). This high efficiency is largely due to elasticity of the thorax -- 90-95% of the potential energy absorbed by flexion of the exoskeleton is released as kinetic energy during the wing's downstroke.
- The migratory locust, *Schistocerca gregaria*, can fly for up to 9 hours without stopping. Large swarms occasionally traverse the Mediterranean Sea. In North America, annual migrations of monarch butterflies (*Danaus plexippus*) wing their way from summer feeding grounds to overwintering sites in California and Mexico. One individual, marked in Ontario, Canada was collected four months later in Mexico, 2800 km from its birthplace.
- Large hawker dragonflies (family Ashnidae) have been clocked at a top speed of 58 km/hr (36 mph) over level ground.
- The wings of a large insect can generate a considerable amount of lift. Green darner dragonflies (*Anax junius*) are able to fly while carrying a load up to 15 times their body weight. Insects with smaller wings have to work much harder to remain airborne. Some of the tiny biting midges (like *Forcipomyia* spp.) beat their wings over 1000 times per second. A special type of muscle tissue is needed to sustain this rapid rate of contraction.

Insect Flight: Some Amazing Facts

Speed: Slowest of 0.02 km/h - Mosquitoes

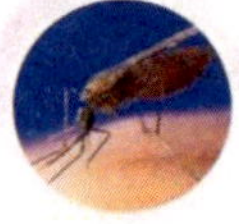

Highest of 146km/h -Horse Flies

Wing Beat : Highest of 1000/ sec - Hover flies

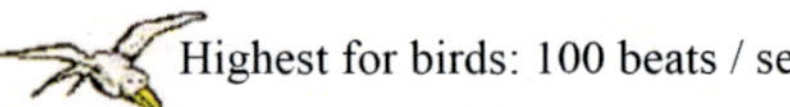

Engineering marvel:
Principles of aerodynamics give no chance for insect flight!!
chance for insect flight!!
But fly they do with uncanny efficiency

Austounding, for they fly with four wings!

Reproductive Potential

i) In insect populations, females often produce large numbers of eggs (high **fecundity**), most of the eggs hatch (high **fertility**), and the life cycle is relatively short (often as little as 2-4 weeks). The queen of an African termite colony may be the mother of more than ten million workers during her 20-25 year lifespan. Queen bee can lay 1500 – 2000 eggs/day. San Jose scale can lay 400-500 eggs/day.

- Since most insects die before they ever have an opportunity to reproduce, a high reproductive potential is the species' best chance for survival.
- Many adaptations help maximize this potential. Most females, for example, can **store sperm for months or years within the spermatheca**, a special region of the reproductive system. A single mating can supply a female with enough sperm to fertilize all the eggs she will produce in her lifetime.
- **An unbalanced sex ratio**, where females outnumber males, is another way to maximize reproductive potential. Since most insects are not monogamous, a few males can supply sperm for a large number of females.
- Many species (e.g. aphids, scale insects, thrips, and midges) where males are entirely absent -- all members of the population are female and contribute offspring through a process of **asexual reproduction**.

Shorter life cycle

Corn root aphid lays 12-16 young ones and reproduction starts from eighth day. A generation is completed in 16 days. Theoretically if all the descendants survive, the chain of these aphids is long enough to encircle the earth. House fly starts its reproduction by April and produces 1,91,010,000,000,000,000 up to August. If 1/8 cubic inch is given to a fly it can cover earth to 47 feet.

Metamorphosis

- Most insects undergo significant developmental changes as they grow from immatures to adults. These changes, collectively known as metamorphosis, may involve physical, biochemical, and/or behavioral alterations that promote survival, dispersal, and reproduction of the species.
- In the more primitive insects, No (Ametabolus) or incomplete metamorphosis (Hemimetabola) the immatures and adults share many characteristics -- they often live in similar habitats and feed on similar types of food.

- More advanced insects (Endopterygotes), however, undergo complete metamorphosis -- a dramatic transformation in form and function between the immature (larval) and adult stages of development. In these insects, a larva is primarily adapted for feeding and growth. Adults differ in feeding habits from larva.
- In the class Insecta, only 9 out of 28 orders undergo complete metamorphosis, yet these 9 orders represent about 86% of all insect species alive today.
- The obvious advantage to this type of development lies in the compartmentalization of the life cycle. Through natural selection, larval form and function can be optimized for growth and feeding without compromising adaptations of the adult for dispersal and reproduction.
- Each stage of the life cycle is entirely free to adapt to its own ecological role.
- A combination of large and diverse populations, high reproductive potential, and relatively short life cycles, has equipped most insects with the genetic resources to adapt quickly in the face of a changing environment.
- Perhaps the most remarkable example of insect adaptation in this century has been the speed with which pest populations have developed resistance to a broad range of chemical and biological insecticides (**Insecticides resistance, Biotypes**).

The ways of growing : Metamorphosis

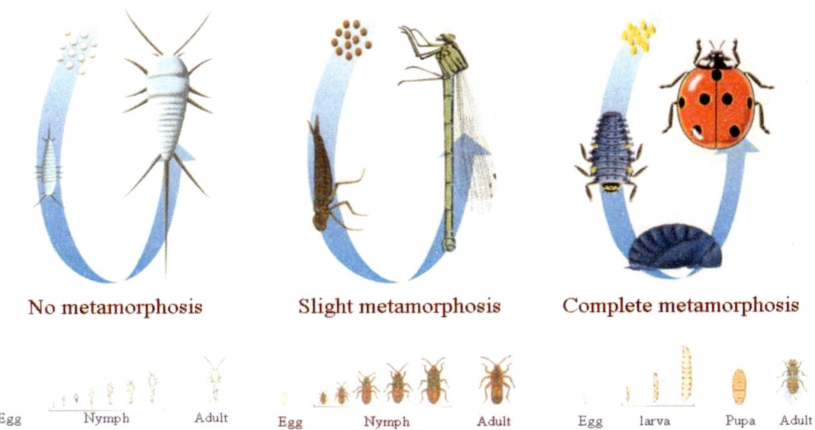

Metamorphosis confers several advantages

Adaptability

Presence of diapause

The development is supressed to tide over adverse conditions. *e.g.* red hairy caterpillar *Amsacta albistriga* under go diapause as pupa.

Defense mechanism

Insects have different types of defense mechanisms.

i) Behavioural defense: It is a defense strategy adopted by some insects through feigning death or imitating the voice of dangerous insects or mimicry. **Colarado potato beetles** when disturbed , draw their legs beneath and drop to the ground and pretend as if dead. This phenomenon is called *thanotosis*. Some insects pupate in soil.

ii) Structural defense: Beetles have hardened forewing (elytra) to protect from predatory birds.

iii) Chemical defense: Some insects produce or release poisonous or unpleasant odours from their body or possess warning coloration by imitating certain distasteful insects.

- Bees and wasps produce venom to defend themselves.
- **Stink bugs** have specialized exocrine glands located in the thorax or abdomen that produce foul smelling hydrocarbons.
- Larvae of **swallow tail butterflies** have eversible glands called **osmeteria,** located just behind the head when disturbed they release repellent volatile and waves their body back and forth to ward of intruders
- Some **blister beetles** (Meloidae) produce **cantharidin**, a strong irritant and blistering agent.

iv) Colouration defense: The body color and shape of some insects make them look like part of the plant, thereby protecting themselves from natural enemies. Colour of the insect resemble with the Surrounding environment. *e.g.* leaf insect, stick insect.

v) Construction of protective structures: Some insects construct shelter with the available plant material for protecting themselves from adverse conditions, natural enemies and to store food material for use during the period of scarcity. **Eg**: Cases / Bags in case of **case worms/bag worms** Termatoria in case of **termites**, Honey comb in case of **honey bees**

Physiological adaptations

i) Cellulose digestion in termites through symbiotic association with bacteria and protozoa.

ii) Rhythms: periodically repeated variation in activity. e.*g*. Circadian rhythm.

iii) Communication system is well developed in social insects like bees, termites and ants.

Pioneers in Chemical Warfare

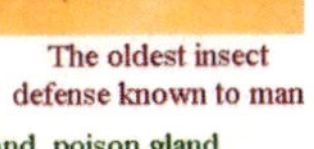

Honey bee

The oldest insect defense known to man

Sting and poison gland

The awful osmeterium of the swallowtail caterpillar

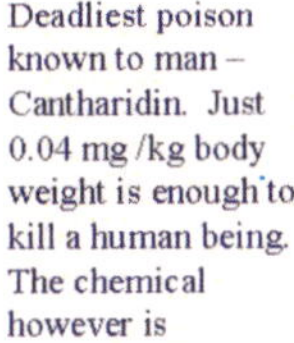

Deadliest poison known to man – Cantharidin. Just 0.04 mg /kg body weight is enough to kill a human being. The chemical however is known for its medicinal value

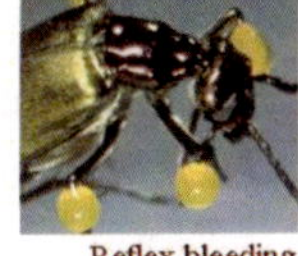

Reflex bleeding in a spanishfly

Defense spray of the bombardier beetle

The Great Escape!

A burning hot jet of gas sprayable in any direction and at high speeds of 500 pulse per second.
How does the insect manage to keep it inside without being hurt?

Painted poisons and Startling threats

Bright colours warn predators- we are poisonous

I collect poison from my food

Don't eat me or you will be sorry!

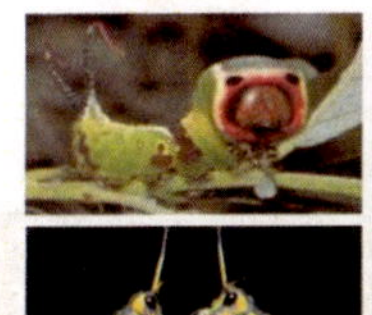

Deception by invitation

These insects startle an approaching predator by suddenly exposing the eyespots

Appearance and disappearance of eyespots frightens the predator away

Predators quickly heed the warnings and learn to avoid them

2

Major Phyla of Animals

American Ecologist Whittaker (1969) proposed the five kingdom system of classification. In his 'Five Kingdom System', he added three more kingdoms in to Linnean model of two kingdom system. These were: Monera, Protista and Fungi. Whittaker succeeded to overcome the difficulties as well as demerits of two, three and four kingdom systems and to represent the living organisms according to the evolutionary relationships among themselves. He also defined the kingdoms by a number of special characteristics such as whether the organisms possessed a true nucleus or not. Whittaker's five kingdom system of classification is based on (a) mode of nutrition (b) cell structure and complexity (c) phylogenetic relationship (d) body organization and (e) reproduction.

The large branches within the Animal Kingdom evolutionary tree are 'phyla' (the singular being 'phylum'). 'Animal phyla' outlines thirty three phyla, although there are still disputes and disagreements concerning how many phyla are needed to classify the less well-known species. The most easily recognised are Porifera, Cnidaria, Arthropoda, Nematoda, Annelida, Mollusca, Platyhelminthes, Echinodermata, and Chordata. Animal phyla each comprise a branch of the evolutionary tree and these branches are always connected to other branches. Because of this, some phyla are more closely related to each other than they are to others.

Sl.No.	Phylum	Approximate known species	Examples
1	Porifera	4500	Sponges
2	Platyhelminthes	12, 700	Flatworms: Planaria, tapeworm
3	Nematoda	12, 000	Roundworms
4	Mollusca	1, 00,000	Snails, octopus, oysters
5	Annelida	7,000	earthworms, leech
6	Arthropoda	1,000,000	Crabs, ticks, mites, shrimps, scorpions, spiders and insects
7	Echinodermata	6500	Starfishes, sea urchins, sand dollars
8	Chordata	43000	Fishes, amphibians, reptiles, birds and mammals
9	Coelenterate	80,000	Corals, Jellyfish, hydra

Distinguishing Characters of Phylum Arthropoda

Animal kingdom comprises of more than 1.75 million kind of animal, which vary is their shape, size and structure. On the basis of similarities and differences in morphological characters animal kingdom is classified into 29 separate groups called as phyla. Amongst various phyla, Arthropoda constitutes the largest single group containing about 1.5 million species.

- Animals are triploblastic and bilaterally symmetrical.
- Body is segmented and jointed externally.
- Body is covered with hardened and nonliving chitinons exoskeleton called as or body wall.
- A variable number of segments carry paired and jointed appendages.
- Animals are coelomates and the body cavity is called as haemocoel.
- Tubular alimentary canal with mouth and anus at anterior and posterior ends.
- The circulatory system is of open type with dorsal blood vessel carrying paired ostia.
- The central nervous system consists of a brain and ventral never cord.
- Striated muscles (with dark and light bands).
- The respiration is with the help of tracheae, or through the body surface or through gills or book lungs.
- The sexes are usually separate and male and female animals are easily distinguishable.
- They usually under go metamorphosis.

Classes of Arthropoda

It is the largest phylum in the animal kingdom. Besides insects, many creatures like crayfish, crabs, lobsters, centipedes, millipedes, spiders, mites, ticks, scorpions etc. come under this category. Phylum arthropoda is classified into 7 classes viz :

- Onychophora (claw bearing): eg: peripatus
- Crustacea (crusta-shell): eg: prawns, crabs, wood louse
- Arachnida (Arachine-spider): eg:scorpion, spider, ticks, mites
- Symphyla
- Myriapoda

- Chilopoda (chilo-lip,poda-legs): eg:centipedes
- Diplopoda (diplo-two,poda-legs):eg:millipedes

- Trilobita (an extinct group)
- Hexapoda is subphylum (hexa-Six; poda-legs) eg. insects
- Insecta (in-internal ;sect-cut)

1. Class - Onychophora eg. peripatus

- They form connecting link between phylum Annelida and Arthropoda
- They are worm like and covered with soft unsegmented exoskeleton (No chitin)
- Pair of ringed antenna
- The excretory organs are **Nephredia**. Salivary glands are present.
- Respiratory – tracheal system has a non segmental arrangement and the irregularly placed stigmata open in to respiratory / tracheal pits.
- They posses 15 to 43 pairs of legs which are unjointed but terminated into claws as in other arthropods.
- Animal are terrestrial

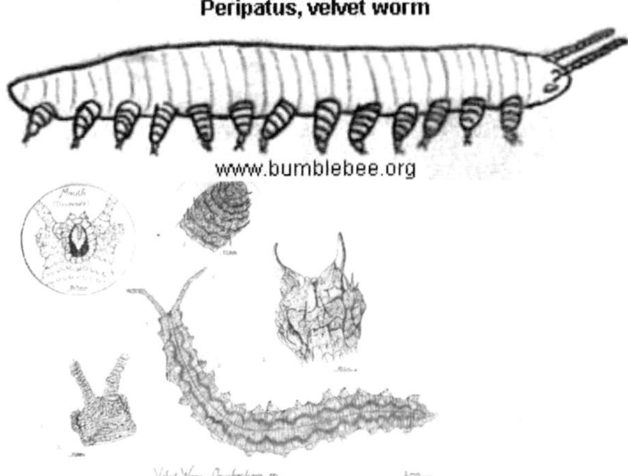

2. Class - Trilobita -eg Trilobites.

- The name trilobite comes from the Greek words *tri* meaning three, and *lobita* meaning lobed. The name refers to the three distinct longitudinal regions of the trilobite body.

- This includes marine forms with the body molded longitudinally into three lobes (unsegmented head, flexible trunk and unsegmented tail.)
- Head bears single pair of antennae followed by number of pairs of limbs.
- Four pairs of limb are attached to head and remaining pairs to trunk region.
- Though they only remain as fossils, the marine creatures called trilobites filled the seas during the Paleozoic era. Today, these ancient arthropods are found in abundance in Cambrian rocks.

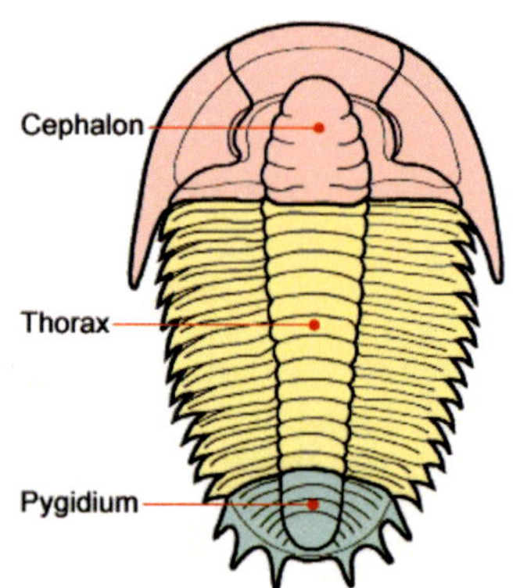

3. Class - Crustecea eg. Prawns, crabs, lobster, Cyclops, woodlice, frillbugs, sowbugs, crayfish, shrimps, barnacles.

- This class includes predominantly marine arthropods in which exoskeleton is generally formed due to the deposition of calcium salt.
- Body is divided into cephalothorax and abdomen.
- They posses five pairs of legs and two pairs of antennae
- Respiration takes place by mean of gills.
- The excretory organs are modified and represented by green glands which lies in head region and open at the base of antennae.
- Reproductive organs are open at the base of legs.
- Sexes separate
- Genital opening in pairs

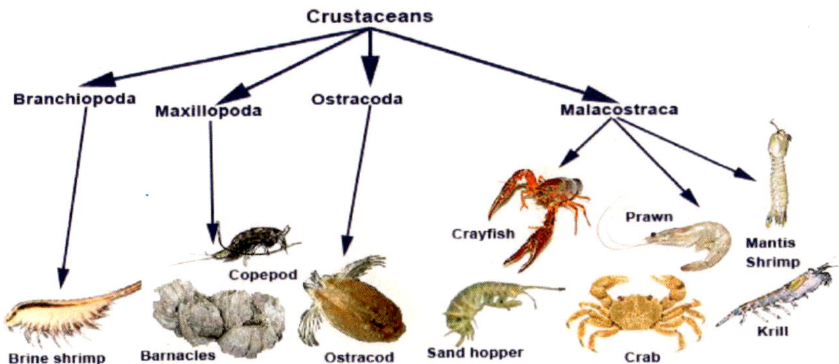

4. Class – Arachnida-eg scorpion, spiders, ticks, mites

- Body is divided into cephalothorax (Prosoma), abdomen (Mesosoma) and tail (metasoma)
- They bear four pairs of legs.
- Antennae and compound eyes are absent instead of antennae they posses chelicerae and pedipalpi.
- Respiration takes place with book lungs or tracheae
- The excretory organs are eventually malphigian tubules.
- Green glands open near the base of abdomen.

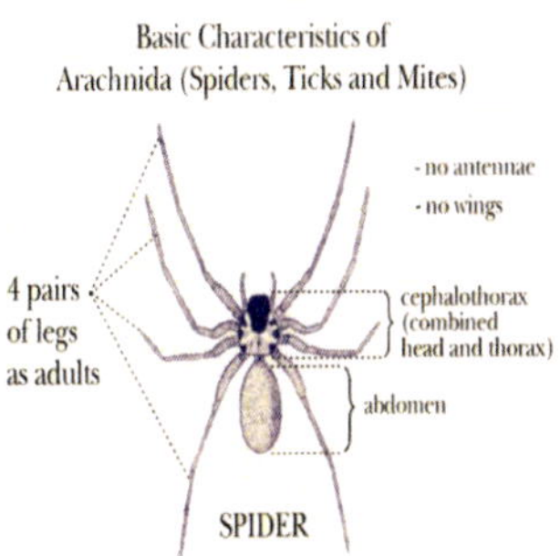

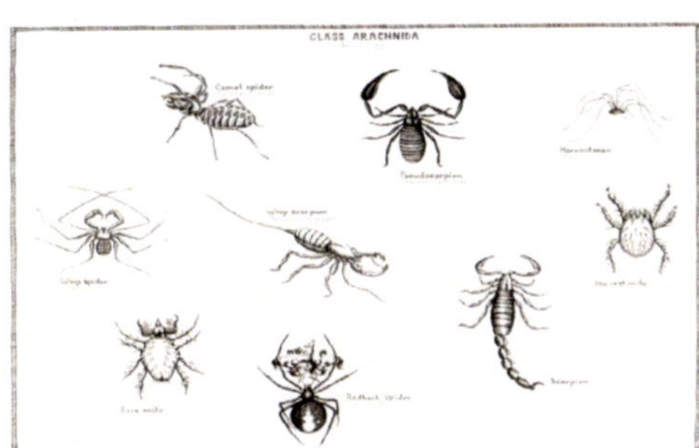

5. Class - Myriapoda -eg centipedes and milliepdes

- Body is divided into head and trunk.
- Head : 6 segmented
- They have single pair of antennae
- Respiration takes place with the help of tracheae.

- Each body segment bears two pairs of appendages (legs)
- 2 sub classes – Diplopoda and Chilopoda

Characterstics	**Diplopoda**	**Chilopoda**
Antenna	Paired, 7-segmented	Paired, many segmented
Body form	Cylindrical, curve in to a close spiral when disturbed	Flattened body
Legs	2 pairs/body segment	1 pair/body segment, the 1st pair directed forward to act as poson claws
Gonads opening	Behind 2nd pair of legs (Progoneata)	Open behind on the penultimate segment (Opisthogoneata)
Spiracles	two pairs of spiracles/body segment	One pair of spiracles/body segment
Members	Millipedes	Centipedes

Millipede

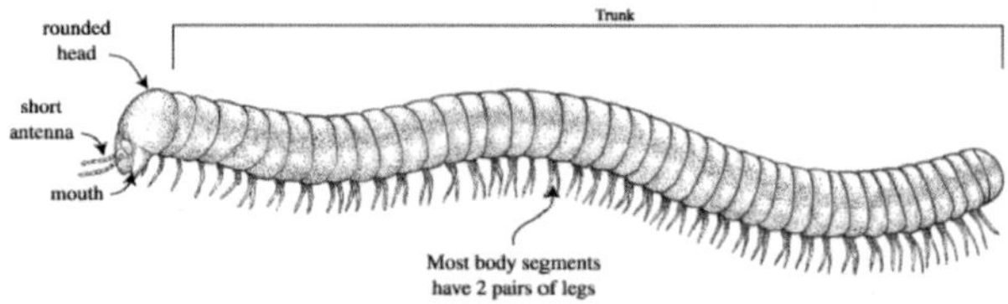

Centipede

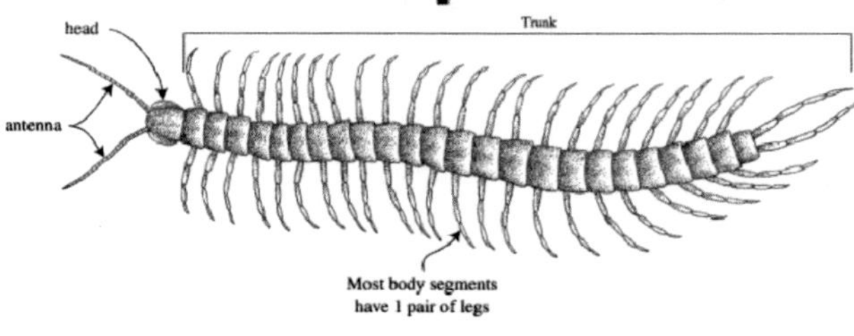

6. Class - Symphyla

- Body 14 segmented
- Each segment has with a pair of appendages, usually a pair of stylets and a reversible appendage on the abdomen.
- Antenna many segmented
- Gonopore on the 4th abdominal segment

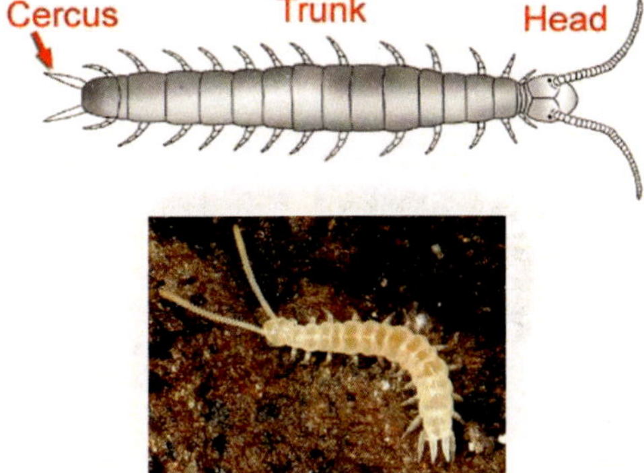

7. Class - Insecta eg. Cockroach, Grasshopper etc.

- The body is divisible into three distinct region head, thorax and abdomen.
- Body is bilaterally symmetrical and triploblastic
- Head bears single pair of antennae.
- Thorax consist of three segment each of which bears a pair of legs and 2nd and 3rd segment carry a pair of wings.
- Abdomen consisting of 11 segments. It is devoid of any ambulatory appendages and genital openings are situated on 8th and 9th abdominal segments.
- Mouth opening situated is in the head where as anus opens at the caudal end of the body.
- The alimentary canal is differentiated into foregut (stomodeum), mid-gut (Mesenteron) and hind gut (Proctodeum).
- The central nervous system composed of brain and sub-oesophageal ganglion and ventral nerve card.
- Circulatory system is open type. The dorsal vessel consists of heart and aorta. The heart is divided into different chambers, each chamber having paired lateral opening called as ostia.
- The respiratory system consists of air tubes or tracheae which communicate with the exterior by means of paired spiracles.
- Excretion takes place by the malphigian tubules.
- Sexes are separate.
- Metamorphosis of often present.
- Sense organs include simple and compound eyes, auditory, olfactory, tactile and gustatory organs, light and sound producing organs may also be present.
- Insect are only invertebrate capable of flight.

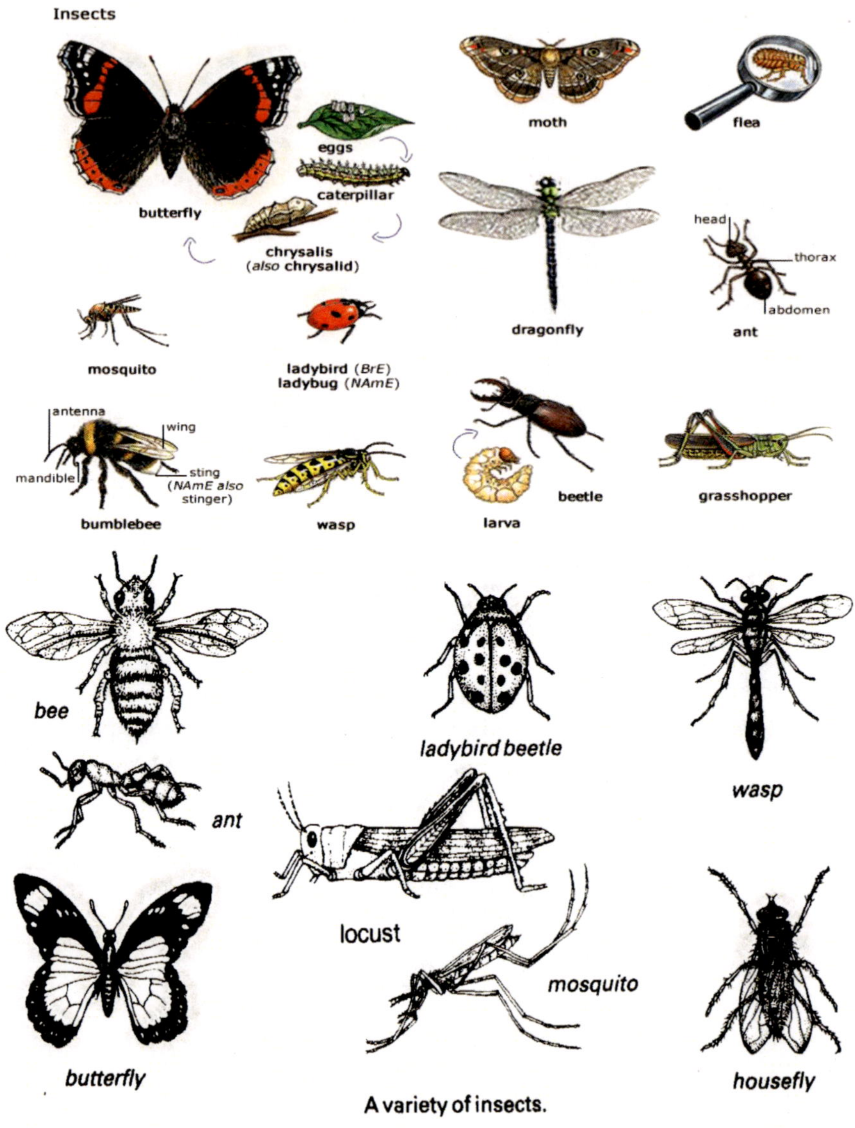

A variety of insects.

Relationship of Class Insecta with Other Classes

Character	**Crustacea**	**Arachnida**	**Myriapoda**		**Onychophora**	**Insecta**
			Chilopoda	**Diplopoda**		
Habitat	Aquatic and Terrestrial	Terrestrial	Terrestrial	Terrestrial	Terrestrial	Many Terrestrial and a few aquatic
Habit	Hervibores, carnivorous	Hervibores, predators	Carnivorous	Hervibores,	Organic matter	Hervibores, predators and parasitiods
Body regions	2 (cephalothorax and abdomen)	3 (Pro, meso and meta soma)	2 (Head and multi segmented trunk)	2 (Head and multi segmented trunk)	Worm like, unsegmented in adults	3 (Head, thorax and abdomen)
Mouthpart	One pair of Mandibles, two pairs of maxillae with maxillary pads	Mouth parts are Chelicerae and pedipalpi	1 pair of mandible and 2 pairs of maxillae	1 pair of mandible and 1 pair of maxillae	Mouth parts are called jaws or oral papillae	Labrum, a pair Mandibles, ap of Maxillae, labiu and hypopharnyx
Antenna	2 pairs	Absent	1 pair	1 pair	1 pair	1 pair
Visual organs	1 pair of stalked compound eyes	1 pair of simple eye	1 pair of simple eye	1 pair of simple eye	1 pair of simple eye	Compound eyes and simple eye
Locomotor organs	5 pairs of biramous appendages	4 pairs	1 pair/segment; 1st pair of leg is modified to form poison	2 pair/segment	Many bilateral lobe like legs	3 pairs of thoracic legs and 2 pairs of wings
Respiration	Gills	Book lungs and trachea	Tracheal	Tracheal	Tracheal	Tracheal
Excretion	Coelomoducts or green glands	Maplhigian Tubules	Maplhigian Tubules	Maplhigian Tubules	coelomoducts	Maplhigian Tubules
Character	**Crustacea**	**Arachnida**	**Myriapoda**		**Onychophora**	**Insecta**
			Chilopoda	**Diplopoda**		

Development	Anamorphosis	Metamorphosis absent (Scorpion); metamorphosis present (mites)	Metamorphosis	Metamorphosis	Anamorphosis	Metamorphosis
Special features	Classification strengthens exoskeleton	Life cycle : Egg-Larva (3 pairs of leg)-Nymph (4 pairs of leg))-Adult	Opisthogenital-gonopore present in the terminal segment	Progogenital-gonopore present in the 3rd segment	Link between Annelida and Arthropoda	Genital structure on 8th and 9th abdominal segment
Examples	Crayfish, crabs, woodlice, lobsters	Spiders, scorpions, ticks, mites	centipedes	Millipedes,	*Peripatus*	Insects

3

General Body Organization of Insects

Body Segmentation

In general, insect body is divided in to a series of segments, which in primitive arthropods are known as "**somites**" or "**metameres**". During the process of evolution, these somites gets fused with each other in different ways forming the body parts of the existing arthropods. The type of arrangement of these body segments in **embryonic stage** is known as **primary segmentation** while in **adult insects** is known as the **secondary segmentation** which differ from primary in having a sclerotized membranous **intersegmental** region.

In adult and nymphal insects, and hexapods in general, one of the most striking external features is the amalgamation of segments into functional units. This process of **tagmosis** has given rise to the familiar **tagmata** (regions) of **head**, **thorax**, and **abdomen**. In this process the 20 original segments have been divided into an embryologically detectable six-segmented head, three-segmented thorax, and 11-segmented abdomen (plus primitively the telson), although varying degrees of fusion mean that the full complement is never visible.

The grouping of body segments in to regions is known as **tagmosis.**

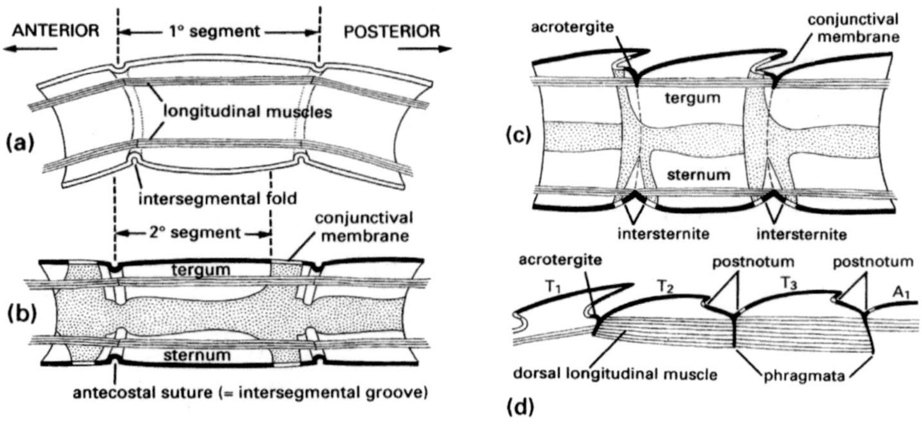

Types of body segmentation.

(a) Primary segmentation, as seen in soft-bodied larvae of some insects. (b) Simple secondary segmentation. (c) More derived secondary segmentation. (d) Longitudinal section of dorsum of the thorax of winged insects, in which the acrotergites of the second and third segments have enlarged to become the postnota. (After Snodgrass 1935)

Head consists of mouthparts, compound eyes, simple eyes (ocelli) and a pair of antennae.

Thorax consists of 3 segments i.e. prothorax, mesothorax and metathorax, Meso and metathorax are together known as **pterothorax**. All the three thoracic segments possess a pair of legs and meso and meta thorax possess one pair of wings.

Abdomen has 7-11 segments with genital appendages on 8th and 9th segments.

Integument

- Insect body is differentiated into three distinct regions called head, thorax and abdomen (grouping of body segments into distinct regions is known as tagmosis
- Body regions are called as tagmata
- Insect body is covered with protective covering called integument.
- The insect body consists of series of ring like structures called segments or somites or metamere.
- The flexible, infolded portion of the cuticle which is present in between the two definite adjacent segments is called as intersegmental membrane.
- The body segments of insects are connected one behind the other, which is called as metameric segmentation.
- Sclerotized or hard part of the cuticle at localized area is called as sclerite, thinner, weaker, membranous, unsclerotized area of the cuticle where actually the body wall breaks at the time of moulting is called as suture.
- Each segment is having four surface regions.
    - Upper or dorsal region is known as dorsum/tergum/Notum in case of thorax.
    - Lower or ventral region is known as sternum/venter.
    - Two lateral regions called as pleurae, each known as pleuron.
- Sclerites composing tergum are known as tergites.
- Sclerites composing sternum are known as sternites.

- Sclerites composing pleuron are known as pleurites.
- The detached plates between two adjacent definite segments are known as intersegmentalia.
- According to the position of intersgmentalia, they are known as –
    - Intertergites
    - Intersternites
    - Interpleurites
- The insect body consists of total 20 to 21 segments divided into head thorax and abdomen.

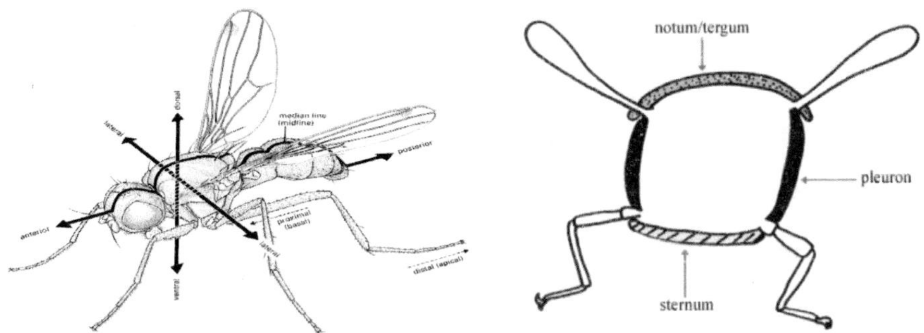

Bodywall Structure

Integument

- The integument is the outer layer of the insect comprising epidermis and cuticle. The cuticle is a characteristic feature of arthropods and is responsible for the success of insects as terrestrial animal.
- Insect integument can be divided in to three basic parts *viz.*,
    - The basement membrane (non-cellular layer)
    - The epidermis (cellular layer)
    - The cuticle (non-cellular layer)

Basement membrane

- It appears as amorphous granular layer (0.1-0.15μm thick). It contains mucopolysaccharide. It is a continuous layer secreted by the blood cells (haemocytes).

Epidermis

- Outer layer one cell thick with cell density ranging from $3000mm^{-2}$ (Trachea of *Rhodinus*) to $11{,}000mm^{-2}$ (sternal area of *Tenebrio*). The apical plasma membrane of epidermal cells forms projections known as plasma membrane plaques which are the sites of secretion of outer cuticle and of chitin fibre.
- The epidermal cells are held together near their apices by zonulae adherens and in lower region down by septate junctions.
- Gap junctions between epidermal cells provide a pathway for movement of hormones etc. and maintain coordination between cells.
- The basal plasma membrane is flat and attached to basal lamina by hemidesmosomes.
- All the epidermal cells are glandular as they secrete cuticle and enzymes concerned in its production and its digestion at the time of molting.

Functions of epidermis

- It secretes the cuticle and basement membrane
- It produces the moulting fluid which dissolves endocuticle before the insect moults
- It absorbs the digested products of old cuticle
- It repairs the wounds
- It determines the surface pattern of insect

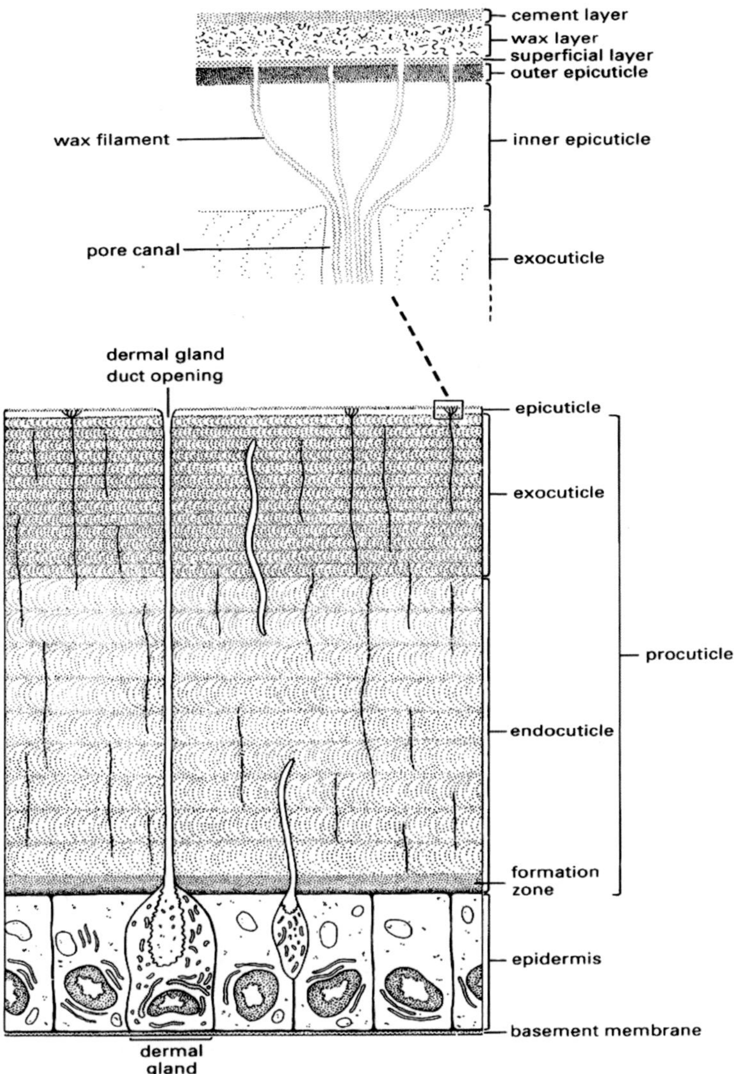

Fig. 1: A typical insect cuticle

Cuticle

- It is a secretion of epidermis and covers the whole of outside of the body as well as lining ectodermal invaginations as stomodaeum and proctodeum and tracheae.
- Differentiated in to 2 regions :
    - Outer thin **epicuticle**, 1-4 µm thick contains no chitins and

  - Inner region **Procuticle** up to 200 μm thick, characterised by the presence of chitin.

- **Epicuticle**
  - The epicuticle is made up of several layers. The thickest layer 0.5-2.0μm thick is the inner epicuticle outside the procuticle. The outer epicuticle is 15nm thick and outside this a wax layer followed by **cement layer**. Chitin is absent. Cement layer is secreted by dermal glands and is composed of lipoproteins. The cement layer is formed through secretions of specialized dermal glands (Verson glands of the epidermal layer), and consists of carbohydrates (lactose), proteins, long chain lipids and some polyphenolic substances. The thin **cement layer/Tectocuticle** consists of mucopolysacchride associated closely with lipids. It is not produced by all insects and absent in honeybees cuticle. It may be folded in various ways and may bear small spines or microtrichia.
  - polyphenol layer is a very thin trilaminar layer, 12-18 nm in thickness. It is a highly polymerized lipid (polyphenol layer) and has a protein component. The material forming the outer epicuticle is referred as **cuticulin.** It is highly resistant to acids and organic solvents.
  - The epicuticular **wax layer** contains 9% hydrocarbons having chain length of around 12 to 50 carbon atoms and is about 0.25 μm thickness. Wax is an important in waterproofing the cuticle and in some insects is the source chemical signals important in intra- and inter specific signalling. The epidermis secretes the lipids of the wax layer just before ecdysis, which is transported via pore canals, and the finer wax canals, and released to the surface. The wax canals are considered the route of entry into the insect for oil-soluble compounds.
- **Procuticle**
- **Exocuticle**
  - Contribute rigidity or toughness to the cuticle and consists mainly of chitin and proteins.
  - It is tanned (pigmented by a hard, brown material)
  - Chitin absent/reduced in intersegments.
  - The **horny nature** of exocuticle is not due to chitin but due to presence of an insoluble material comprising tanned proteins k/such as **Sclerotin.**

- **Sclerotisation /hardness** of the cuticle or sclerotisation is due to the interaction of proteins with tanning agents involving various phenolic substances.
- **Tanning** in adult cuticle is caused by a peptide hormone **Bursicon** produced by the Neurosecretory cells (NSC) of the brain and released form the thoracic-abdominal ganglia. Bursicon was first detected in Blowflies.

- **Endocuticle**
  - Thickest layer of procuticle made up of horizontal lamellae.
  - Contain chitin and proteins but this layer is not tanned.
  - In addition two layer Meso cuticle and Sub cuticle also sometimes exist. **Mesocuticle** lies between exocuticle and endocuticle and is differentiated on the basis of staining reaction. Granular **Subcuticule** is the newly secreted endocuticle that has not been reached its final structural configuration.
- **Pore canals**
  - Very fine pore canals of 1 μm or less in diameter runs through the cuticle at right angles extending from epidermis to inner epicuticle.
  - The density as high as 15000mm^{-2}.
  - These are concerned with the transport of lipids, oxidising enzymes used for sclerotisation of cuticle and material for wound repair from epidermis to the surface of cuticle.

Composition of Cuticle

- **Chitin**
  - Main component of procuticle comprising 20-50% of its dry weight $(C_8H_{13}O_5N)x$. It is a polysaccharide made up of N-acetylglucosamine, the sugar residues linked by 1-4 β linkages. The chitin microfibrils are about 2.5-3.0nm in diameter and are embedded in a protein matrix. The arrangement of microfibrils may be parallel, heliciodal, or pseudo-orthogonal etc.
  - It is insoluble in water, alcohol, organic solvents, dilute acids and concentrated alkalies.
  - Hardening of cuticle is due to the cross linking of protein molecules (tanning or sclerotization) to form a rigid matrix.

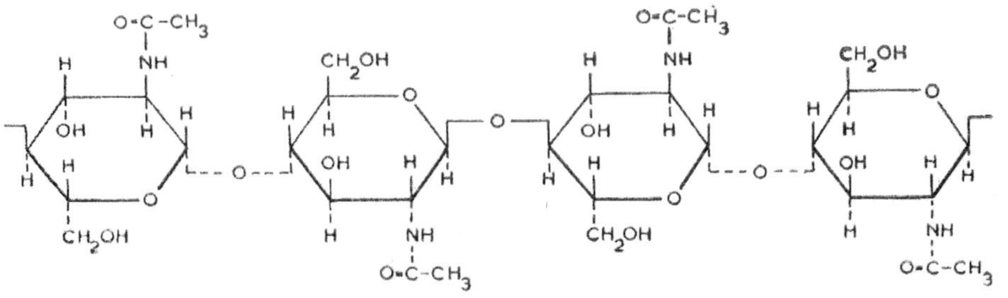

Structural formula of chitin

- **Arthropodin**
  - Non chitinised water soluble substances of cuticle (25-37%) of the dry weight of insect cuticle. There are three different types of protein.
  - Arthopodin: It is the water soluble fraction of insect protein.
  - Sclerotin: It is alkaline soluble. The thorny nature exocuticle is due to sclerotin.
  - Resilin: It is rubber like and is present in wing articulation and thorax, acting like mechanical springs.

Function of Cuticle

Support

- Determine the form and size of insect
- Support of internal organ
- Muscle attachment
- Form different organs inside the body e.g. Tentorium, apodemes etc.
- The tubular, external skeleton of the legs provides great strength and relative lightness compared with internal skeleton of the vertebrates.
- The presence of hard jointed appendages makes accurate movements possible with a minimum of muscles and by lifting the body of the ground facilitates rapid movement.
- The rigidity of cuticle forming wings also makes flight possible.
- The cuticle is modified into various structures that can receive stimuli.
- The epidermal cells that secrete cuticle also secretes and deposits various chemicals such as Pheromones and cuticular pigments within or on cuticle

that are involved in behavioural sequences, mating and evading from enemies due to colour changes / mimics.

Protection

- Heavy sclerotized cuticle in some insects make it difficult for predators to catch or parasitization by parasitoids. Protection from external environment is also afforded. It also prevent movement of pathogens, parasite, and dangerous chemicals into the body
- Presence of thick cuticle on the abdomen prevents abrasion by the substratum.
- The Cuticular lining of the fore gut and hind gut also protects the epidermis from abrasion by the food.

Barrier to the water loss

- Wax layer present on the cuticle prevent water loss and plays a major role in the success of insects as some terrestrial organisms.

Movement

- The rigidity that the exoskeleton exhibits makes it possible for muscle movements precisely, due to insertion of muscles to the integument wall tightly. Wing movement is possible only because of hard cuticular flight Sclerites in thoracic region.

Insect Cuticular Outgrowths

Suture-a seam or impressed line indicating the division of the distinct parts of the body wall. It is a groove marking the line of fusion between two distinct plates. This makes no reference to primary segmentation or the presence of internal apodemes.

Sulcus-a furrow or groove, a groove-like excavation. It is only a ridge giving strength against strain imposed on the head capsule.

Two types of processes formed by the cuticle :

Internal : apodemes, tentorium

External : setae, spurs, spines, hairs, tubercles. These are characterised by the absence of membranous articulations.

Cuticular Structures

Non-cellular

Non cellular appendages have no epidermal association , but rigidly attached. Eg. minute hairs and thorns.

Cellular

Cellular appendages have epidermal association and it may be unicellular, multicellular.

a. Unicellular structures:

- Clothing hairs, plumose hairs. e.g. Honey bee.
- Bristles. e.g.flies.
- Scales - flattened out growth of body wall e.g. Moths and butterflies
- Glandular seta. e.g. caterpillar
- Sensory setae - associated with sensory neuron or neurons
- **Seta** - hair like out growth from epidermis. Epidermal cell generating seta is known as **Trichogen**, while the socket forming cell housing trichogen is known as **Tormogen**. Study of arrangement of seta is known as **Chaetotaxy**.

b. Multicellular structures: e.g. Spur – movable ; Spine- immovable.

Cuticular Appendages

Setae

Modification of epidermis. These are hollow outgrowths of the entire bodywall lined by a layer of epidermal cells. They may be of various shape but essentially they have a membranous articulations. The base of seta is set in a small setal membrane, depressed in a hair socket or alveolus. The epidermal cell forming a seta is called trichogen. Closely associated it is another cell forming the setal membrane which forms the floor of the socket is called the tormogen. The arrangement of more constantly located seta is called chaetotaxy.

Types of setae

i) *Clothing hairs*: They are present on the surface of the body or appendages. When the clothing hairs have thread like branches they are termed as plumose hairs. *e.g.* honey bees. When the setae are stout and rigid they are called bristles. *e.g.* Tachinid flies.

ii*)* *Scales*: Scales are modified clothing hairs which occur in butterflies and moths

iv) Glandular seta: When the seta functions as the outlet for epidermal glands it is called as glandular seta. When the glandular seta is stout and rigid it is called as glandular bristle. *e.g.* Urticating hairs of caterpillars

v) *Sensory seta*: This seta is sensory in function and it is connected to nervous system.

Spur

Spurs are multi cellular in origin. Movable (because there is a socket). There may be more than one trichogen cell to form spur. Spurs usually occur on legs of insects. *e.g.* honey bee middle leg.

Spurs	**Spines**
Cuticular appendages	Cuticular processes
Movable, multi cellular structure and thick walled	These are movable outgrowths of cuticle
eg. present on tibia of plant hopper and honeybees	Eg. Hind tibia of grasshopper and leaf hoppers

Apodemes	**Apophysis**
Hollow cuticular invaginations which provide area for muscle attachment	Solid invaginations of the cuticle which gives mechanical support to various organs by forming distinct skeletal structures.

Cuticular Processes

Cuticular processes are outgrowths rigidly connected to the cuticle without any membranous articulation. The arrangement of bristles on an exoskeleton of an insect is called chaetotaxy. The chaetotaxy is being studied for taxonomic purposes.

Types of cuticular processes:

i) *Microtrichia*: They are called fixed pairs of aculei (aculees). They occur as minute structures on the wing of flies.

ii) *Spine*: Spines are outgrowths that are thorn like. It is seen in dung rollers. Not movable

iii) *Nodules*: Nodules are conical outgrowths.

iv) *Tubercles:* A small knob like or rounded protuberance.

v) *Horns*: Horns are very large projections.

vi) Multicellular Spine

Outgrowth of integument (epidermis & cuticle) present on the legs of many insects which are of multicellular origin.

Dermal Glands

Wax glands	Honeybee (4th -7th abdominal sterna segments), scale insects, mealy bugs
Lac glands	Lac insect
Repugnatorial glands	plant bugs (6th and 7th abdominal segments)
Odoriferous glands	Rice gundhi bug nymphs(5th Abdominal segment)
Pygidial glands	Beetles
Androconia (scent glands)	on Wings of some male lepidopteran
Hypodermal poison glands	associated with setae or spines in Slug caterpillar, Saturniid moths;cause irritations

Molting

The insect cuticle is hard and forms unstretchable exoskeleton and it must be shed from time to time to permit the insects to increase their size during growth period. Before the old cuticle is shed new one has to be formed underneath it . This process is known as moulting

Moulting is a complex process which involve

1) **Apolysis** 2. **Ecdysis** 3) **Sclerotization**

Apolysis : [Apo = formation ; Lysis = dissolution]

The dissolution of old cuticle and formation of new one is known as apolysis. Apolysis starts with repeated mitotic division of epidermal cells resulting in increase in number and size of epidermis, which becomes columnar in shape and remain closely packed. Because of this change, the epidermal cells exerts tension on cuticular surface and as a result get separated themselves from the cuticle. Due to separation of epidermis from the cuticle a sub cuticular space is created and the epidermal cells starts producing their secretion i.e. moulting fluid and cuticular material into this space. The moulting fluid is granular, gelatinous and contains two enzymes viz., proteinase and chitinase which can dissolve the old cuticle. As the moulting fluid digest the old cuticle, the sub cuticular space increases gradually by the same time and is occupied by the newly formed cuticular layer, the polyphenol layer, wax layer and cement layer into the deposition of definite layers of epicuticle. Procuticle get deposited beneath the epicuticle and subcuticular space is fully occupied. Though moulting fluid is capable of digesting the entire endocuticle, some undigested old exo and epicuticle portions will remain as a layer in the form of an ecdysial membrane.

Ecdysis

The stage where the insect has both newly formed epi and procuticle and old exo and epicuticle is known as pharate instar. The ecdysial membrane starts splitting along the line of weakness due to muscular activity of the inner developing insect and also because of swallowing of air & water resulting in the distention of the gut.

The breaking at the ecdysial membrane is also due to the pumping of blood from abdomen to thorax through muscular activity. After the breakage of old cuticles which is known as exuviae, the new instar comes out bringing its head followed by thorax, abdomen and appendages.

Sclerotization

After shedding of old cuticle the new cuticle which is soft, milky white coloured becomes dark and hard through the process known as tanning (or) sclerotization. The process of hardening involves the development of cross links between protein chains which is also known as sclerotization. This tanning involves the differentiation of procuticle in to outer hard exocuticle and inner soft endocuticle.

Steps In Molting

- Behaviroual changes: Larva stops feeding and become inactive.
- Changes in epidermis: In the epidermis cell size, its activity, protein content and enzyme level increases.Cells divide miotically and increases the tension, which results in loosening of cells of cuticle.
- Apolysis -- separation of old exoskeleton from epidermis
- Formation of sub Cuticular space
- Secretion of inactive molting fluid by epidermis
- Production of cuticulin layer for new exoskeleton
- Activation of molting fluid
- Digestion and absorption of old endocuticle
- Epidermis secretes new procuticle
- Ecdysis -- shedding the old exo- and epicuticle
- Expansion of new integument
- Tanning -- sclerotization of new exocuticle

Ecdysis: Periodical process of shedding the old cuticle accompanied by the formation of new cuticle is known as moulting or ecdysis. The cuticular parts discarded during moulting is known as **exuvia**. Moulting occurs many times in an insect during the immatured stages before attaining the adult-hood. The time interval between the two subsequent moulting is called as **stadium** and the form assumed by the insect in any stadium is called as **instar**.

An insect that is actively constructing new exoskeleton is said to be in a **pharate** condition. Ecdysis, however, occurs quickly (in minutes to hours). A newly molted insect is soft and largely unpigmented (white or ivory). It is said to be in a **teneral** condition until the process of tanning is completed (usually a day or two).

Control of Moulting

It is controlled by endocrine glands like prothoracic gland which secrete moulting hormone. Endocrine glands are activated by prothoracico-tropic hormones produced by neurosecretory cells of brain.

JH : Juvenile Hormone :Produced from corpora allata of brain that helps the insects to be in immature stage.

MH : Moulting hormone: Produced from prothoracic glands of brain that induces the process of moulting

Eclosion Hormone: Released from neurosecretory cells in the brain that help in the process of ecdysis or eclosion.

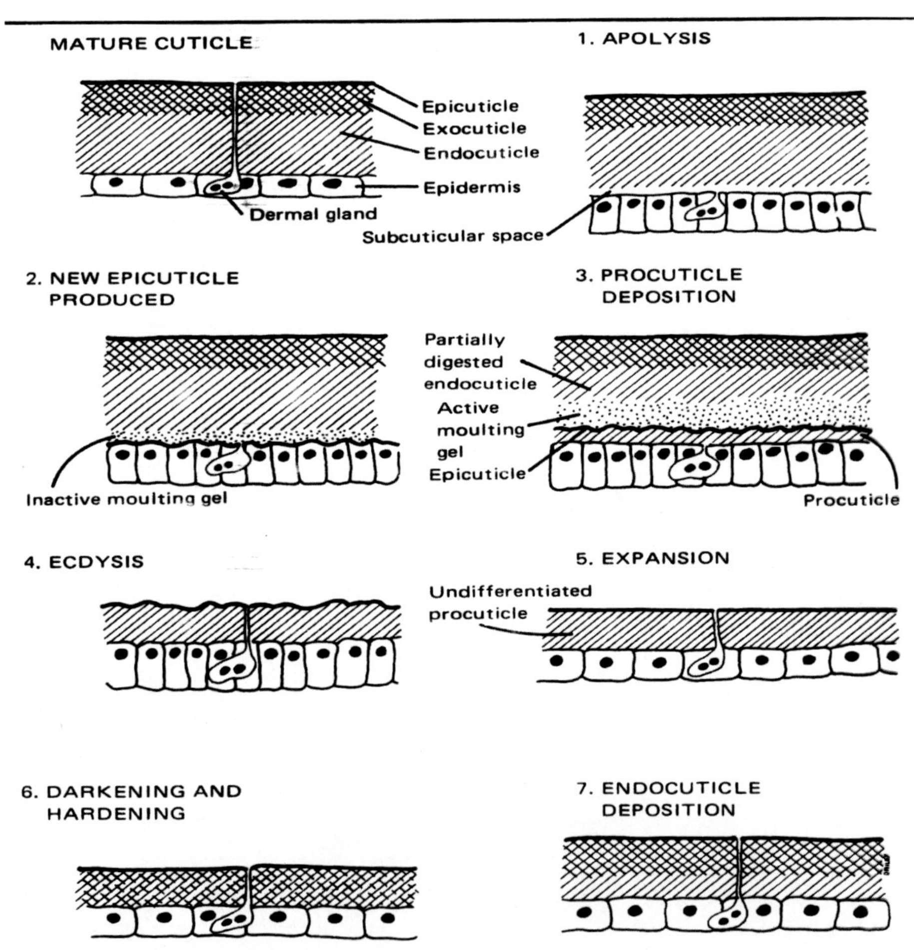

4

Insect Head

Endoskeleton of the Head (Tentorium)

- The tentorium forms during development when 2-4 pairs of apophysis/ apodemes (finger-like invaginations of exoskeleton) fuse internally to create a "bridge".
- The major components of the tentorium are the anterior tentorial arms, the posterior tentorial arms, the dorsal tentorial arms, and the tentorial bridge.
- The anterior tentorial arms originate as invaginations at the lateral limits of the frontoclypeal suture. The coalescence of the anterior tentorial arms forms a more or less solid structure, the tentorial bridge.
- The posterior tentorial arms arise from pits at the ventral ends of the post-occipital suture and they unite to form a bridge running across the head from one side to the other.
- In many insects, e.g. Orhtoptera, the central part of the composite tentorium enlarged to form a broad plate referred to as the **Corporotentorium**.
- The tentorium serves as an internal "truss" that reinforces the head capsule, cradles the brain, provides attachment for muscles and provides a rigid origin for muscles of the mandibles and other mouthparts. The antennal muscles also arise from the tentorium.

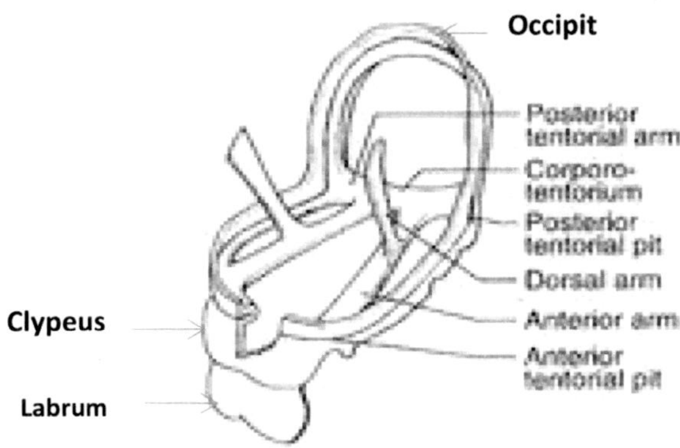

Insect Head

It is the first and anterior most part and the smallest body region. It carries the sense organs i.e. compound eyes and antennae. Brain is present inside the head. It is a hard and highly **sclerotized** compact structure. It is the foremost part in insect body consisting of 6 segments that are fused to form a **head capsule.** The head segments can be divided in to two regions i.e. **procephalon** and **gnathocephalon** (mouth).

1. Procephelon: The part of an insect's head that lies anteriorly to the segment in which the mandibles are located.
2. Gnathocephelon: The part of the insect head lying behind the protocephalon; bears the maxillae and mandibles.

Composition of head

Head is formed by Acron and 6 segments.

Acron is pre-oral lobe like structure observed during embryonic stage. Each of the six segments have paired neuromere and paired appendages, as follows.

Segment	Region	Neuromere	Coelom sac	Appendage
1. Preantennal	Procephelon	Protocerebrum	Sometimes present	Absent
2. Antennal		Deutocerebrum	Usually present	Antennae
3. Premandibular (Intercalary)		Tritocerebrum	Usually present	Embryonic
4. Mandibular	Gnathocephelon	Mandibular ganglion	Usually present	Mandibles
5. Maxillary		Maxillary ganglion	Usually present	Maxillae
6. Libial		Labial ganglion	Usually present	Labium

Mandibular, maxillary and Labial ganglion comprise sub-oesophageal ganglion. The insect's head is sometimes referred to as the head-capsule, and is the insect's feeding and sensory centre. It houses the brain, a mouth opening, mouthparts used for ingestion of food, and major sense organs (including antennae, compound eyes, and ocelli).

Functions of Head

- Food ingestion
- Sensory perception
- Coordination of bodily activities
- Protection of the coordinating centers

Orientation of Insect Head

Insect head is of three types viz., ***hypognathous, prognathous and opisthognathous*** head according to its orientation with respect to the rest of the body.

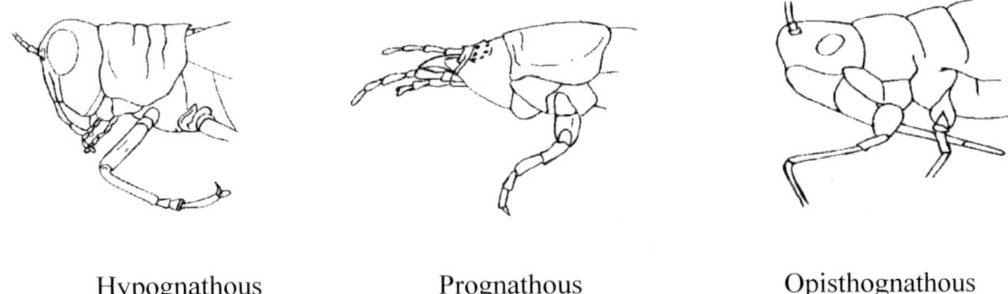

Hypognathous Prognathous Opisthognathous

Hypognathous head (*hypo*, below; *gnathos*, jaw)

The long axis of the head is vertical and the mouthparts are ventral in position. This type can also be called orthopteroid type. *e.g.* grasshopper.

Prognathous head (*pro*, in front of ; *gnathos*, jaw)

The long axis of the head is horizontal and the mouthparts point forward. It is also known as **coleopteroid** type. *e.g.* beetle.

Opisthognathous head (*opistho*, behind; *gnathos*, jaw)

The head is so deflexed that the specialized elongated mouthparts slope backwards between the fore legs; often referred to as **Hemipteroid** (**opisthorhynchous**) type. *e.g.* bugs.

Structure of Head

The surface of the head is divided into regions (sclerites) by a pattern of shallow grooves (sutures). The uppermost sclerite (dorsal surface) of the head capsule is known as the **vertex**. The triangular sclerite that lies between these frontal sutures is called the frons. The insect's neck is known as the **cervix**. This is a membranous area that allows considerable freedom of movement for protraction and retraction of the insect's head

The upper-mid portion of an insects face is called the 'frons' below this is the 'clypeus' and below this the 'labrum' to either side of which may be seen the edges of the 'mandibles' in some insects various aspects of the 'maxilliary' palps may extend beyond and or below these even when viewed from front on.

The rest of the front of the head: that bit which is above the frons is known as the 'vertex'; the sides of the head are known as the 'gena'.

Frons

The area of the face below the top two 'ocelli' and above the 'frontoclypeal sulcus' (if and when this is visible) and in between the two 'frontogenal sulci', it supports the 'pharyngeal dilator' muscles and in immature forms it bears the lower two arms of the ecdysial cleavage lines.

Genae ("cheeks")

These are lateral sclerites that lie behind the frontal sutures on each side of the head. Below each gena there may be another sclerite (the **subgena**), separated from the gena by a subgenal suture A pair of compound eyes, sockets for two antennae and one or more ocelli (simple eyes) also may be found on the front, top, or sides of an insect's head.

Clypeus

The area of the face immediately below the frons (with which it may be fused in the absence of the frontoclypeal sulcus) and the frontoclypeal sulcus. It supports the 'cibarial dilator' muscles and may be divided horizontally into a **'postclypeus'** and **'anteclypeus'**.

Labrum

It is equivalent to the insect's upper lip and is generally moveable, it articulates with the clypeus by means of the 'clypeolabral suture'.

Sclerites in Insects

- **Labrum :** It is small sclerite that forms the upper lip of the mouth cavity. It is freely attached or suspended from the lower margin of the **clypeus**
- **Clypeus:** It is situated above the labrum and is divided in to anterior **ante-clypeus** and posterior **post-clypeus.**
- **Frons :** It is the facial part of the insect consisting of **median ocellus**.
- **Vertex :** It is the top portion of the head behind the frons or the area between the two compound eyes.
- **Epicraniun :** It is the upper part of the head extending from vertex to occipital suture.
- **Occiput :** It is an inverted "U" shaped structure representing the area between the epicranium and post occiput.
- **Post occiput :** It is the extreme posterior part of the insect head that remains before the neck region.
- **Gena :** It is the area extending from below the compound eyes to just above the mandibles
- **Occular sclerites :** These are cuticular ring like structures present around each compound eye

- **Antennal sclerites :** These form the basis for the antennae and present around the scape which are well developed in Plecoptera (stone flies)

All the above sclerites gets attached through cuticular ridges or sutures to provide the attachment for the muscles inside.

Common sutures present in head

- **Clypeolabral suture :** It is the suture present between clypeus and labrum. It remains in the lower margin of the clypeus from which the labrum hangs down.
- **Clypeofrontal suture or epistomal suture:** The suture present between clypeus and frons
- **Epicranial suture:** It is an inverted 'Y' shaped suture distributed above the facial region extending up to the epicranial part of the head. It consists of two arms called **frontal suture** occupying the frons and stem called as **coronal suture.**
- This epicranial suture is also known as **line of weakness** or **ecdysial suture** because the exuvial membrane splits along this suture during the process of ecdysis.
- **Occipital suture:** It is 'U' shaped or horseshoe shaped suture between epicranium and occiput.
- **Post occipital suture:** It is the **only real suture** in insect head. Posterior end of the head is marked by the post occipital suture to which the sclerites are attached. As this suture separates the head from the neck, hence named as real suture.
- **Genal suture:** It is the sutures present on the lateral side of the head i.e. gena.
- **Occular suture:** It is circular suture present around each compound eye.
- **Antennal suture:** It is a marginal depressed ring around the antennal socket.

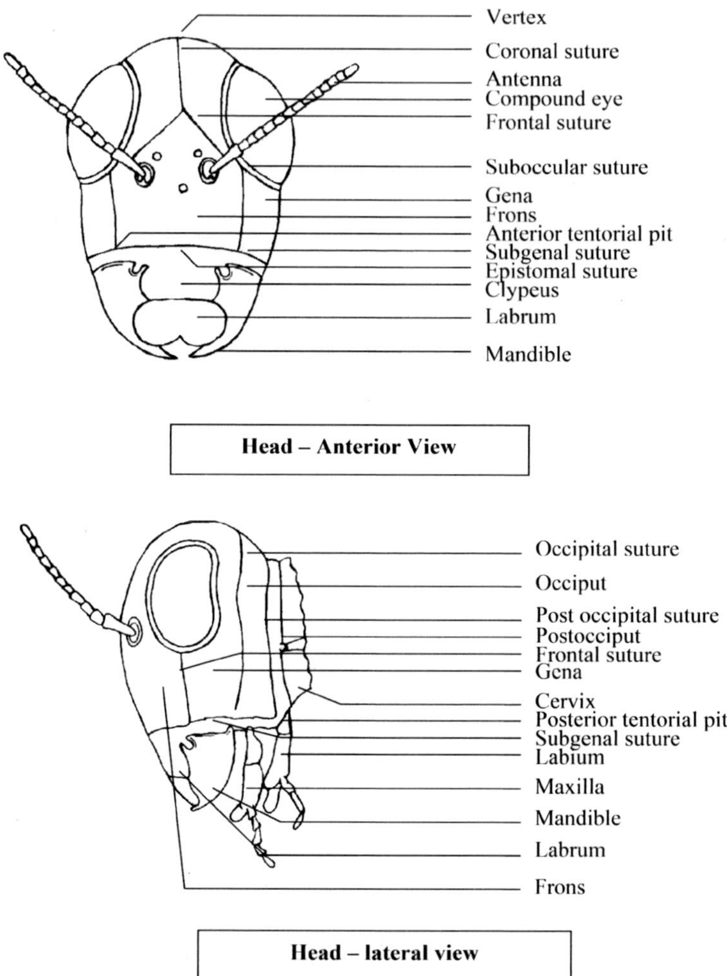

Fig. 1: Head – sutures and regions

Mouthparts : Structure and Modifications

The mouthparts are formed from appendages of all head segments except the second. In omnivorous insects, such as cockroaches, crickets, and earwigs, the mouthparts are of a biting and chewing type (**mandibulate**) and resemble the probable basic design of ancestral pterygote insects more closely than the mouthparts of the majority of modern insects. Extreme modifications of basic mouthpart structure, correlated with feeding specializations, occur in most Lepidoptera, Diptera, Hymenoptera, Hemiptera, and a number of the smaller orders.

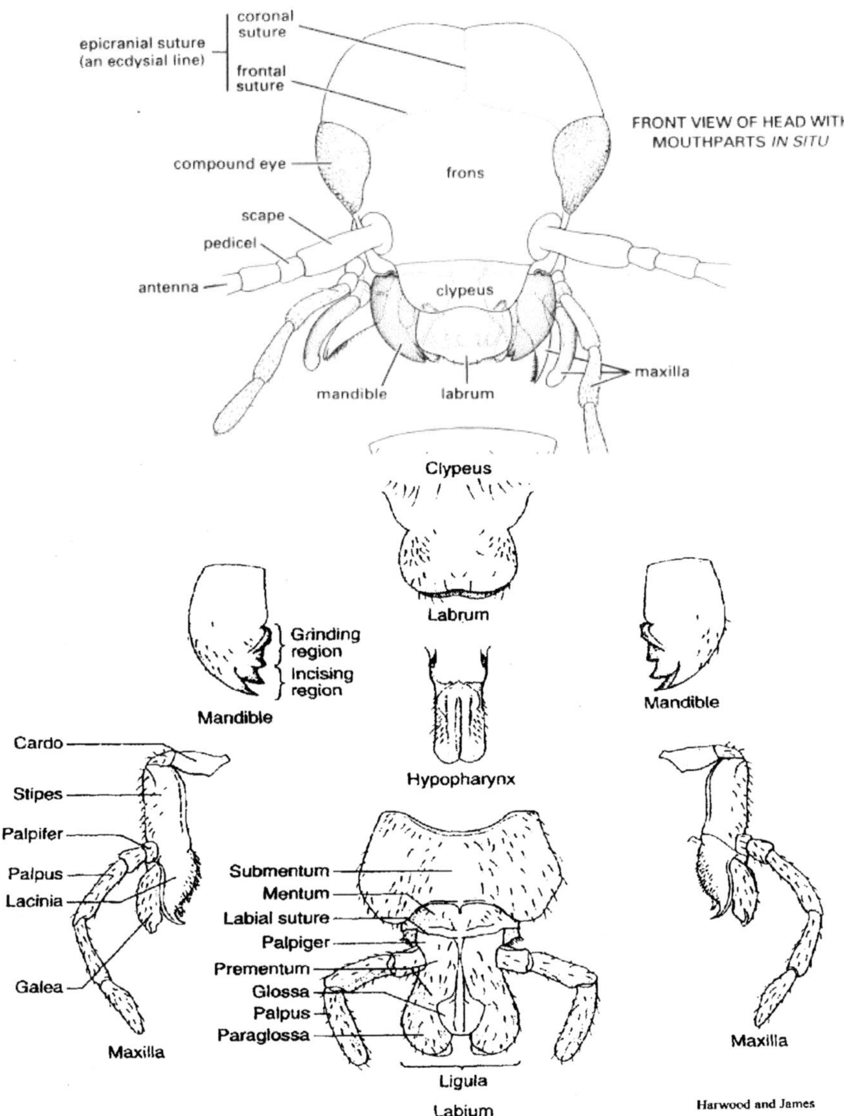

Insect mouthparts typically consist of a labrum, a pair of mandibles, a pair of maxillae, a labium and a hypopharynx. These structures are variously modified in different insect groups. The mouth parts of insects can be basically grouped in to following types based on the type of food and method of feeding.

Type of mouthparts	Examples
Biting and chewing type (Mandibulate)	Dragonflies and damselflies (order Odonata), termites (order Isoptera), adult lacewings (order Neuroptera), beetles (order Coleoptera), ants (order Hymenoptera), cockroaches (order Blattaria), grasshoppers, crickets and katydids (order Orthoptera), caterpillars (order Lepidoptera).
Sucking type / Haustellate type	
• Piercing and sucking type	Cicadas, aphids, and other bugs (order Hemiptera), sucking lice (order Phthiraptera), stable flies and mosquitoes (order Diptera).
• Rasping and sucking type	Thrips
• Sponging type	Adult House flies and blow flies
• Chewing and lapping	Honeybee
• Siphoning type	Butterflies, Moths and skippers
Other type	
• Mask Type	Naiads of Dragonflies
• Degenerate type	Maggots of Diptera
• Mandibulo-suctorial mouthparts	Ant-lion grub

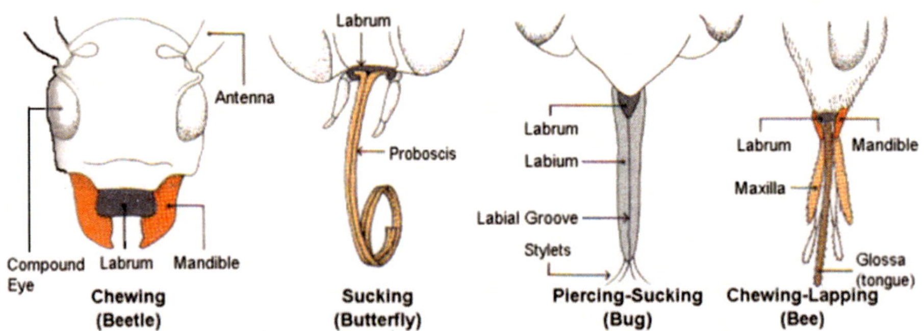

Biting and Chewing Mouthparts (*e.g.*) cockroach, grasshopper.

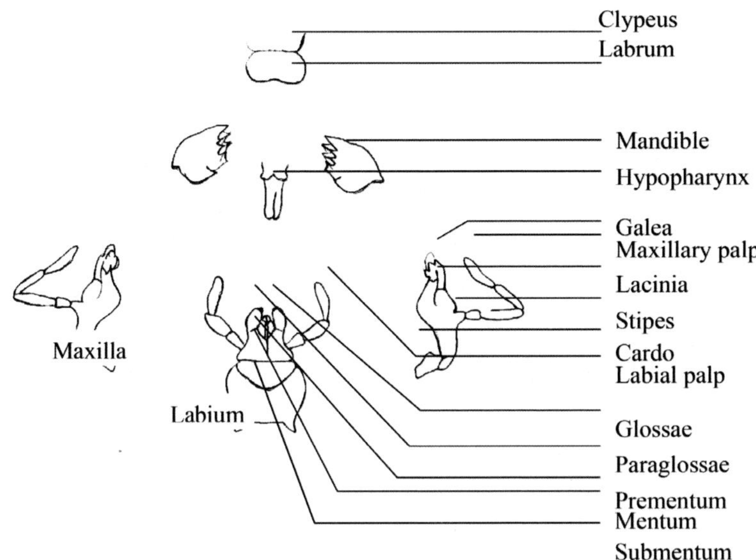

Labrum (*upper lip*)

- Flap-like attached to the *clypeus* overlies the bases of mandibles, forming part of the *pre-oral food cavity*.
- Helps to pull the food into the mouth, close the front of mouth cavity, and to protect mandibles
- On the inner side of the labrum is a sensory lobe, the *labrum-epipharynx* which contains the organs of taste.

Mandible

- Paired heavily sclerotized jaws lying behind the labrum.
- Mandibles have cutting (incisor) and grinding (molar) edges (cusps / teeth) which act like a pair of scissors. Incisors are useful for cutting and molars are useful for grinding. So, the mandibles are adopted for cutting or crushing the food. The mandibles are frequently used for defense in some insects. Each mandible is moved by powerful abductor and adductor muscles.

Maxillae (sr. maxilla)

- Paired present behind the mandibles.
- Have two segments: the basal/ proximal *cardo* (pl. cardines) and the distal *stipes* (pl. stipites).
- Stipes bears two lobes, namely, an inner toothed *lacinia* (pl. laciniae) and an outer *galea* (pl. galeae).
- Sensory *maxillary palp / palpus* (pl. palpi) is commonly five segmented borne on the *palpifer* of stipes.
- Lacinia aids in cutting, grasping or grinding moving similar to the mandibles.

Labium or lower lip

- Comprises a basal *postmentum* and a distal *prementum.*
- Postmentum is divided into the basal *submentum* and the distal *mentum.*
- Prementum carries two pair of lobes, the median *glossae* (sr. glossa) and the lateral *paraglossae* (sr. paraglossa). When these lobes are fused together as in grasshopper and beetles, they are termed the **ligula.**
- The 3 segmented *labial palpi* occur on the *palpiger* of prementum.

Hypopharynx

- Tongue-like median prolongation on the floor of the pre-oral food cavity. In most insects the ducts from the salivary glands open on or near the base of hypopharynx.

Haustellate Mouthparts

Haustellate mouthparts are primarily used for sucking liquids and can be broken down into two subgroups: those that possess stylets and those that do not. **Stylets** are needle-like projections used to penetrate plant and animal tissue. The modified mandibles, maxilla, and hypopharynx form the stylets and the feeding tube. After piercing solid tissue, insects use the modified mouthparts to suck liquids from the host. Some haustellate mouthparts lack stylets. Unable to pierce tissues, these insects must rely on easily accessible food sources such as nectar at the base of a flower. One example of nonstylate mouthparts are the long siphoning proboscis of butterflies and moths (Lepidoptera). Although the method of liquid transport differs from that of the a Lepidopteran proboscis, the rasping-sucking rostrum of some flies are also considered to be haustellate without stylets.

Piercing-Sucking Mouthparts

• BUG TYPE (Hemiptera)

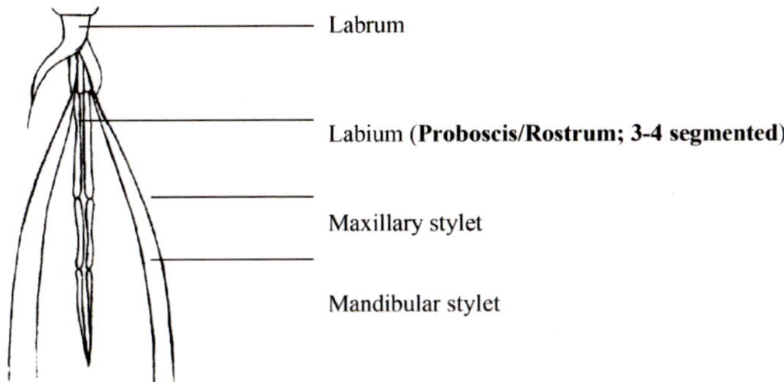

The labium is modified into an elongate, usually segmented sheath-like *beak* (*proboscis*, *rostrum*) which encloses two pairs of piercing organs called *stylets* that are derived from two mandibles and two maxillae. Labrum occurs as a short lobe at the base of beak on the anterior side. Both the palpi are absent. The *maxillary stylets* fit together to form a salivary canal as well as a food canal.

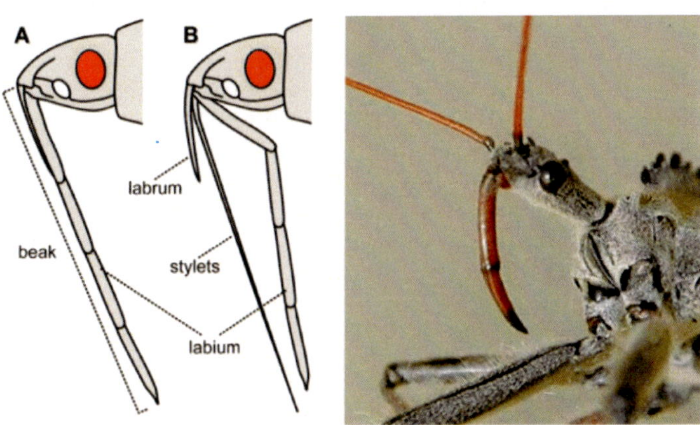

Bug type

• Mosquito Type

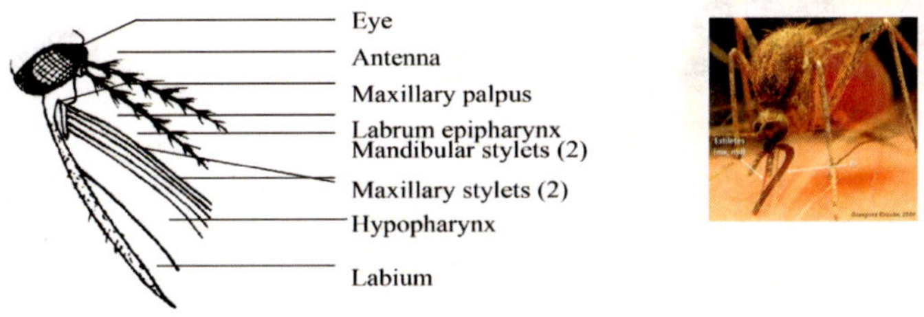

The labial beak encloses six stylets, i.e. two mandibles, two maxillae, one each of labrum - epipharynx and hypopharynx. Maxillary palpi alone are present. The salivary canal runs through the hypoharyngeal stylet while the food canal is formed between the hypopharynx and the overlying slender grooved or U-shaped **labrum-epipharynx**.

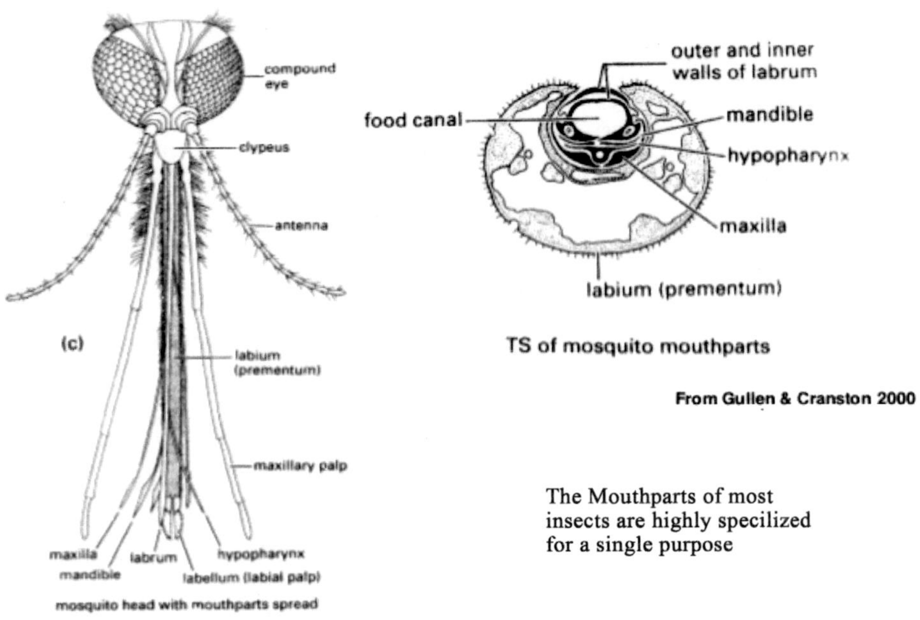

Mosquito type

Mandibulo-Suctorial Mouthparts *e.g.* ant-lion grub

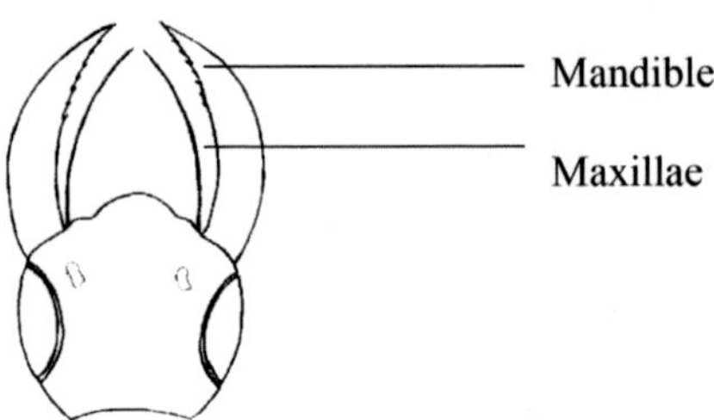

The mandibles are sickle-shaped armed with sharp spiniform teeth. The needle-like maxillae are accommodated against the grooves of the mandibles, forming a canal through which the insect injects salivary fluid and sucks the haemolymph from the crushed prey.

Rasping And Sucking Mouthparts *e.g.* thrips

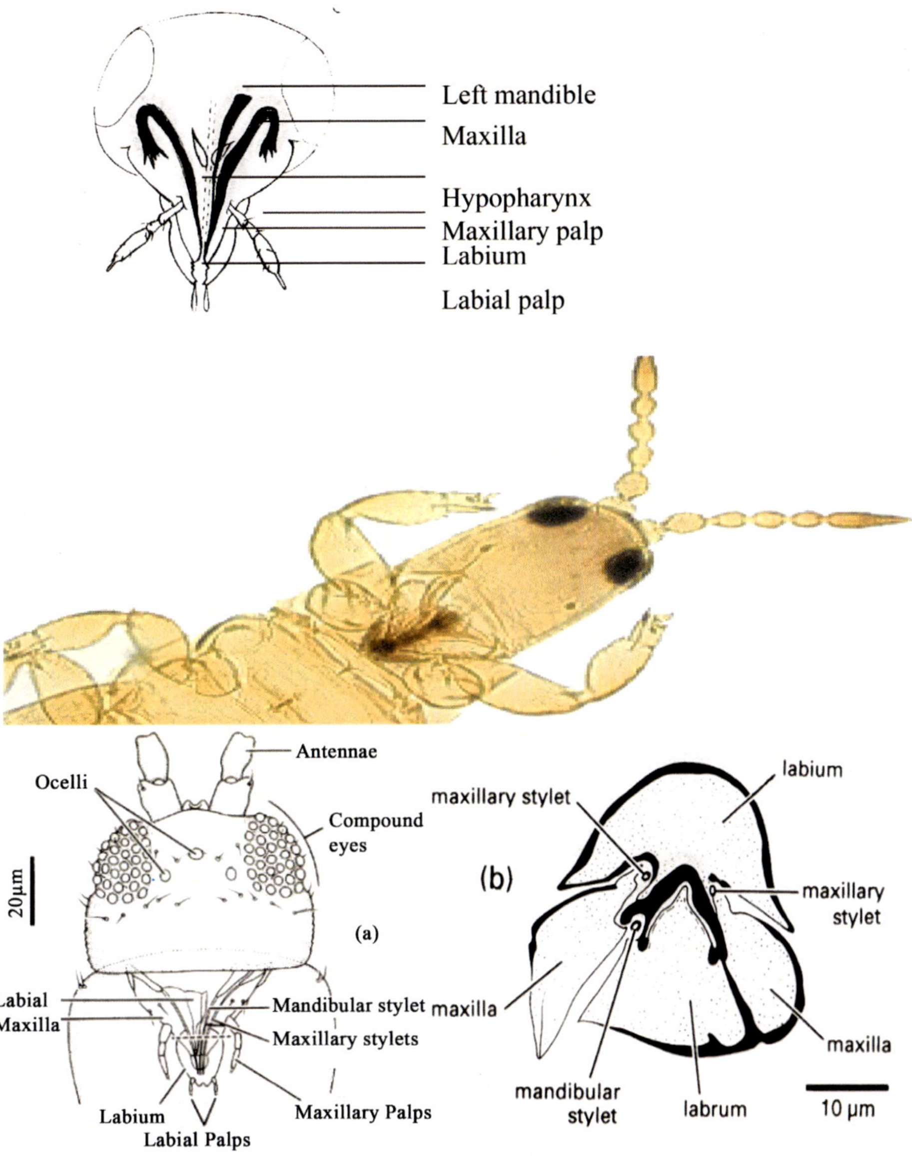

Thrips

The *mouthcone* (rostrum, proboscis), is bound by the labrum in front, the maxillae on the sides and the labium on the back. It encloses only three stylets derived from the left mandible and the two maxillae. **Absence of the right mandible** makes these mouthparts characteristically *asymmetrical*. The hypoharynx is a small median lobe in the proboscis. The maxillary and labial palpi are present. As the *asymmetrical* stylets rasp or lacerate the plant tissues, the exuding plant sap is sucked up by the mouthcone itself.

Siphoning Mouthparts *e.g.* butterflies and moths

The coiled **proboscis** is formed by **the two interlocking maxillary galeae**. The labial palpi are well developed. These insects siphon off the nectar through the proboscis erect from blood pressure. There is no special salivary canal. Mandibles are absent. Labrum not well developed.

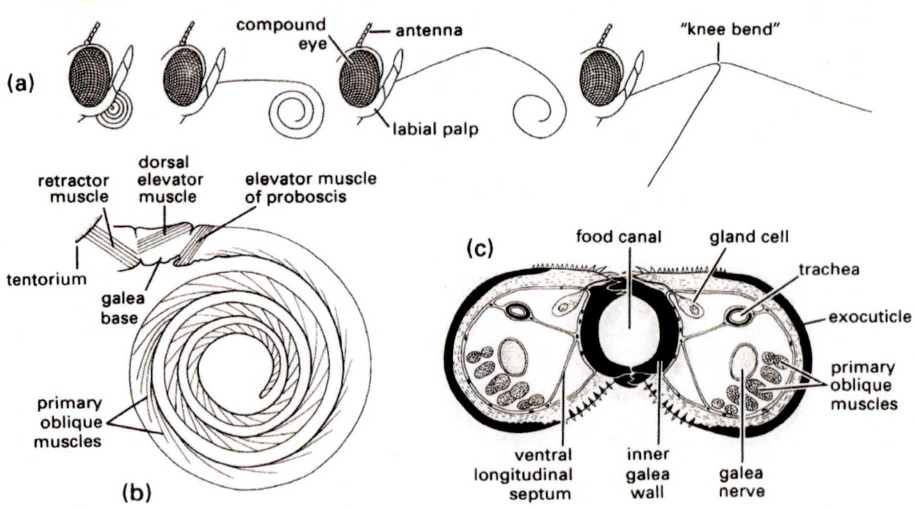

Siphoning type

Sponging Mouthparts *e.g.* (Diptera; non blood sucking flies) house fly

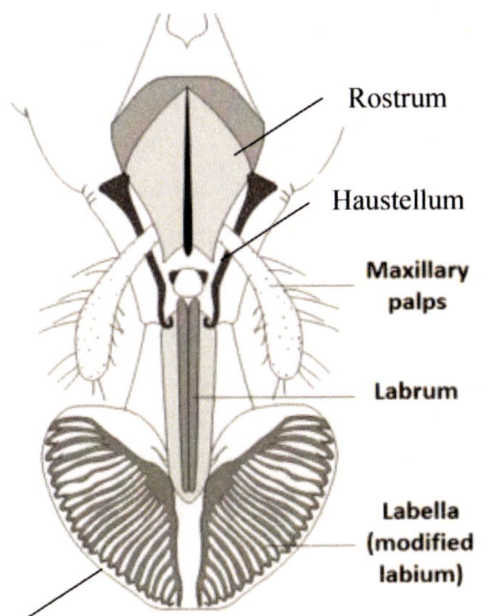

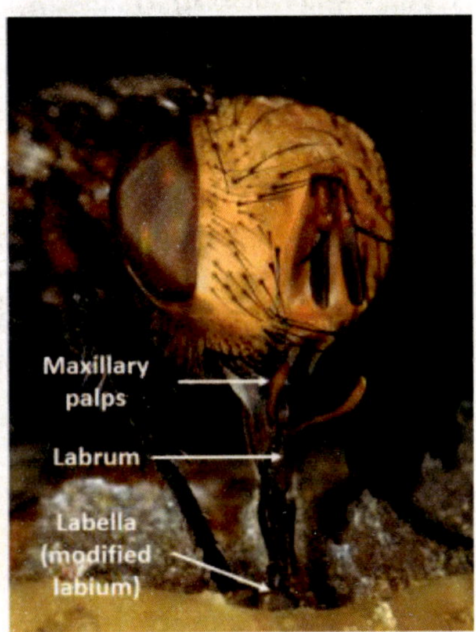

Sponging type

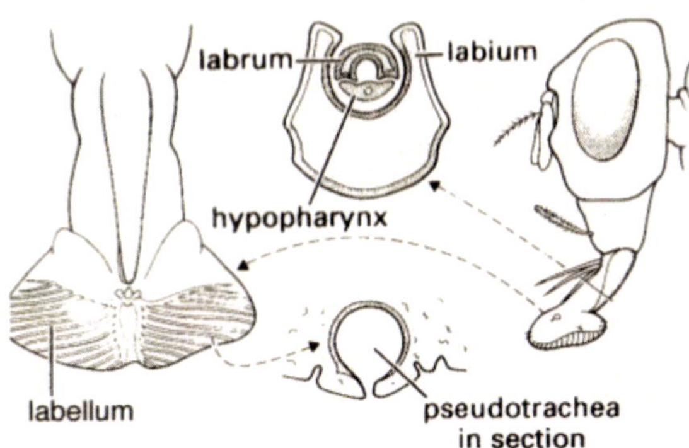

Labium is fully devlopped in to Proboscis.

Rostrum/**basiproboscis** form the basal part of the labium.

Haustellum/**mediproboscis** : distal part of the rostrum to the maxillary palpi is called the Haustellum that ends in a pair of sponging organs, the *labella* (sr. labellum).

Ventrally, the labellum has numberous transverse grooves, the *pseudotracheae.* They converge at one point forming a *median reservoir* or *prestomum*, from which the dissolved liquid food is drawn up through the food canal. The slender labrum and hypopharynx lie in the anterior groove on the labium. The salivary canal is in the hypoharynx. The food canal lies between the labrum-epipharynx and hypoharynx.

Chewing-Lapping Mouthparts *e.g.* bees and wasps

- The labrum and mandibles form the chewing unit. The labrum is a small plate below the clypeus. The mandibles are short and stumpy adapted for biting, chewing and grasping.
- The maxillae and labium are elongate and closely united to form a lapping tongue. Both the pairs of palpi are present, the maxillary palpi being small and the labial palpi elongate.

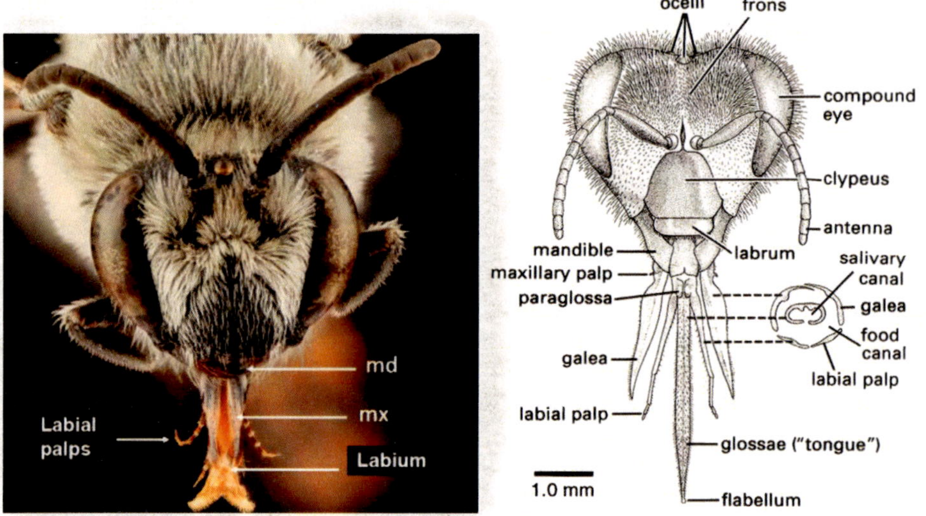

Chewing and Lapping type

- The glossae of the labium are greatly elongate to form a hairy tongue that can be protracted and retracted to reach deep into the nectaries of tubular flowers.
- A temporary food canal is created by the concave inner surface of the maxillary galeae, roofing over the glossa with the labial palpi fitting lengthwise tightly against the sides of the glossa. The salivary canal lies in the glossa.

Mask Type of Mouth Parts : e.g. Naiads of dragon flies.

Mainly useful for catching the prey. **Labium** is modified in to a mask where the prementum and post mentum forms in to an elongated structure with a joint. The labial palpi are represented as teeth like structures / spines at the tip of the labium that are helpful for catching the prey. All other parts remain rudimentary (reduced). During resting period, when the insect is not feeding, the mouthparts cover a part of the head. Hence it is called mask type.

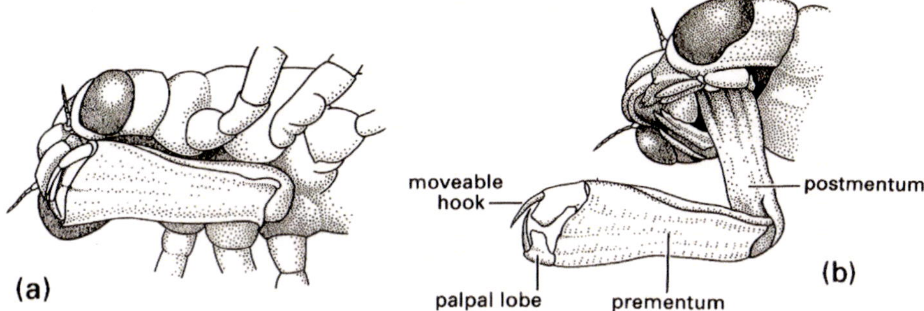

Figures : Dragonfly nymph(Odonata: Aeshnidae: *Aeshna*) a) in folded position, and (b) extended during prey capture with opposing hooks of the palpal lobes forming claw-like pincers. (After Wigglesworth 1964)

Degenerate Type of Mouth Parts : e.g.:Maggots of Diptera.

In apodous maggots a definite head is absent and mouth parts are highly reduced and represented by a mouth hooks/ Spines .

An Extreme Case: The Atrophy

Adult forms of some insects, such as mayflies (Ephemeroptera) or some dipterans, suffer a total reduction of their mouthparts. In these cases, the only function of adults is down to reproduction, so they lose all feeding functions and structures when metamorphose.

Credit : www.agefotostock.com

5

Antennae (Feelers)

Antennae are paired, highly mobile, segmented appendages present between or below or behind the compound eyes. The antennae arise from the 2nd or antennal segment of the head possessing nerves coming from **deutocerebrum** of the brain. Antennae are absent in order **protura** and class **Arachnida** whereas 2 pairs of antenna (antennules) are present in class **Crustacea**.

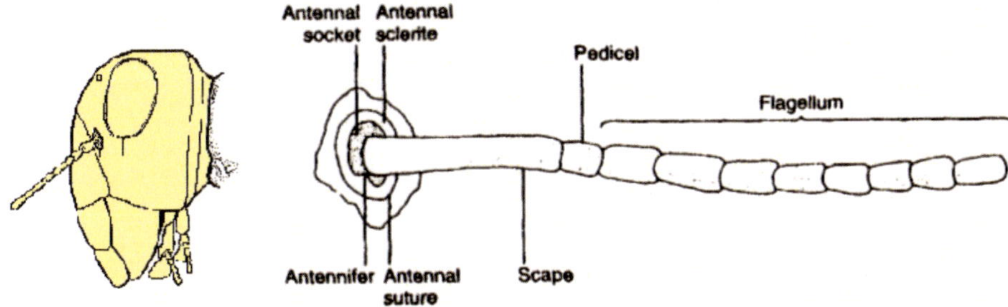

Structure

Antennae are set in a socket of the **cranium** called antennal socket (anetnnifer) and connected by an articulatory membrane which allows free movement of antenna. Antennal socket is provided with an antennal suture. Although antennae vary widely in shape and function, all of them can be divided into three basic parts:

- **Scape**
  - Basal segment, larger than other segments and articulates with the head capsule. It is provided with intrinsic muscles.
- **Pedicel**
  - Second segment of antennae. In this, mass of cells called Johnston **organ** (auditory) is present (eg. houseflies). The above two segments have intrinsic muscles ie., their own muscles. In honey bee, wasps pedicel forms the pivot between scape and flagellum.
- **Flagellum/Clavola**
  - The remaining segments or annuli or flagellomeres are called flagellum which lack individual muscles. Flagellum is supplied with many sensory receptors that are connected by nerves to brain. Flagellum is further divided into three parts

    - Ring joints: It is basal segment of flagellum are small and ring like form.
    - Club: It is swollen or enlarged distal segments of the antenna.
    - Funicle: segments between ring joints and club.
- Flagellum varies in size and shape.

In the family **Chalcidoidea,** the flagellar segments are divided in to the basal ring segments **funicle** and terminal **club.**In general there are no muscles in the flagellum and hence the antennae are called **annulated type**.

In **collembola** and **Diplura**, the flagellar segments are muscular in nature and regarded as true segments and the antennae is known as **segmented type**. **Jhonston's organ** is absent in Collembola & Diplura.

Types of Antenna

The antennae of insects are modified in many ways. Some of these modifications just provide greater surface area for sensory receptors, while others are unique adaptations that bestow special sensory capabilities, such as detecting sound vibrations, wind speed, or humidity. The most common antennal types are listed below:

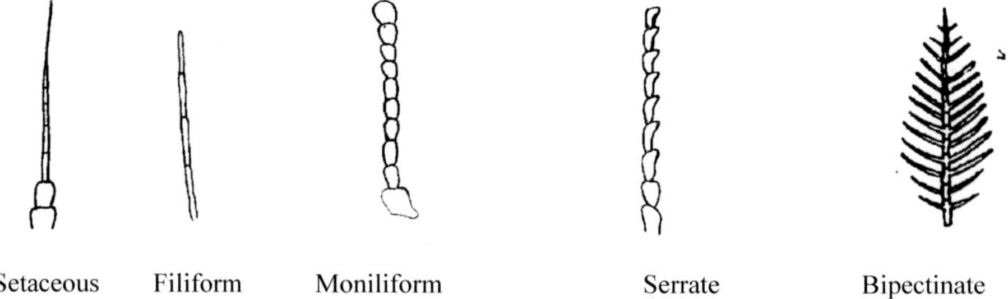

Setaceous Filiform Moniliform Serrate Bipectinate

1. **Setaceous (bristle like)**: The antenna is bristle-like with a marked decrease in the size of the segments from the base to apex. *e.g.* cockroach, dragonfly, damselfly, leafhopper.
2. **Filiform (Thread like)**: Thread-like in form, the segments are nearly uniform in size with no prominent constrictions in between the joints. *e.g.* grasshopper.
3. **Moniliform (beads like)**: This antenna looks *like a string of beads*, the segments being similar in size with conspicuous constrictions between them. *e.g.* termites, midges.
4. **Serrate (saw like)**: Saw-like or saw-toothed, the segments have short, triangular projection on one side. *e.g.* Pulse beetle, click-beetle, Jewel beetle.

5. **Pectinate (comb like)**: Each segment has lateral projections like a comb either on one side (unipectinate e.g Sawfly, Female arctid moth) or more often on two sides (bipectinate, *e.g.* silk worm moth, Male Lymantrid moth).

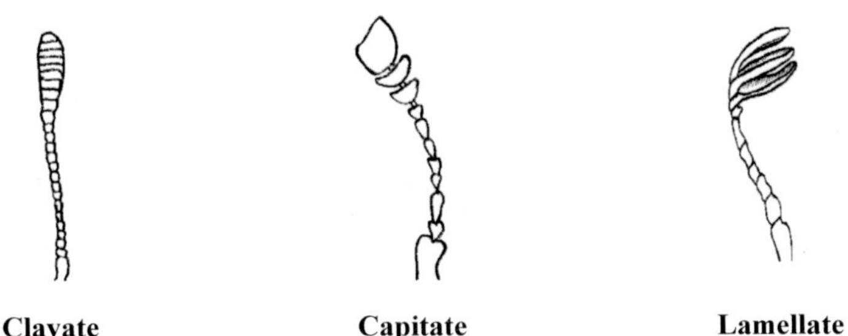

Clavate **Capitate** **Lamellate**

6. Clubbed: The distal segments show an increase in size in the clubbed attenna. Three types of clubbed antenna can be recognised.

a. **Clavate (club like)**: The segments gradually enlarge in size towards the tip. *e.g.* blister beetle, butterflies ending in a club like apical part.

   In Skippers (Lepiodoptera), Segments gradually increase in diameter from base to tip and the last one ends with a small hook like structure.

b. **Capitate (knob like)**: The terminal segments (3-5) alone are enlarged. *e.g.* red flour beetle, butterfly.

c. **Lamellate (plate like)**: The terminal segments expand laterally to form rounded or oval sheath - like lobes. *e.g.* dung-roller, Lamellicorn beetle, Chaffer beetles, Rhinoceros beetle.

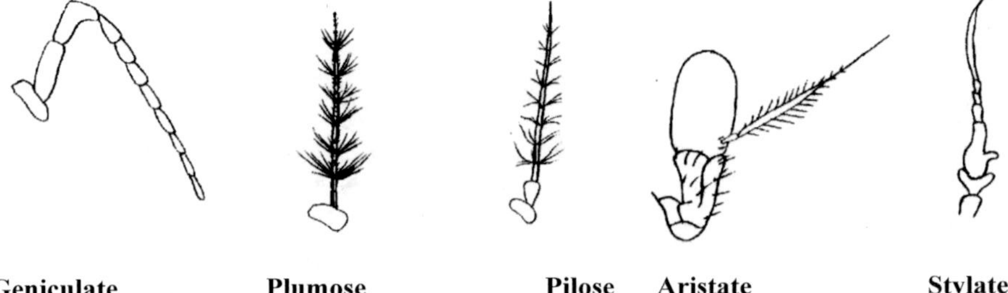

Geniculate **Plumose** **Pilose** **Aristate** **Stylate**

7. **Geniculate (Elbowed/knee like)**: The antenna has a sharp bend like a flexed arm, the first segment being long and at an acute angle to the rest. *e.g.* honey bees, Ants, weevils.

8. **Plumose (Feather like)**: Most flagellar segments have feathery whorls of long hairs. Whorls of hairs arise from each joint of the segment. Each whorl contains number of hairs. *e.g.* male mosquito.

9. **Pilose (Hair like)**: The hairy flagellomeres have a few hairs in whorls. *e.g.* female mosquito.
10. **Aristate**: In this, three segmented antenna the enlarged terminal segment bears a prominent dorsal bristle, the arista. *e.g.* House fly.
11. **Stylate (Bristle like)**: The terminal segment of this antenna carries an elongate, finger-like process, the style. *e.g.* Snipe fly, Robber fly, Horse fly.
12. Flabellate (feather like) : Projections of some upper segments become long and form a feather like structure called **flabella. Eg. Stylopids**

Functions of Antenna

- Perceives sound : Diptera (Male mosquito)
- Measurement of air speed by the chordotonal organ (Johnston's organ) : Orthoptera (Crickets), Coleoptera,.
- Facilitates formation of air of air funnel in water beetle *Hydrophilus*
- Assists mandibles in masticating prey in larva of *Hydrophilus*
- Mating *e.g.* fleas, Collembola, Males of *Meloe* sp.
- Smell : Ants, honeybees, gaint moths
- Taste : Cockroach
- Sexual character : Diptera and Lepidoptera
- Seizing prey : *Chaoborus* sp.

Images : Antennae (Feelers)

Sphingid moth Saturnid moth Silkworm moth

Bipectinate

Male (Plumose)

Female (Pilose)

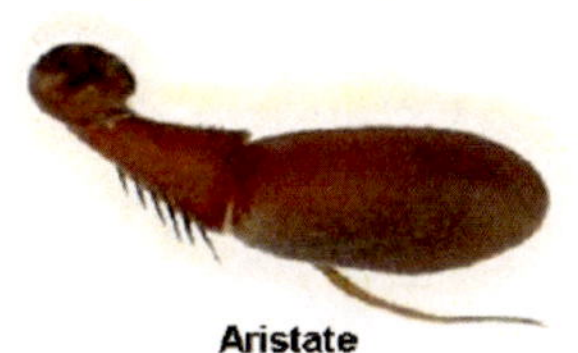

Capitate

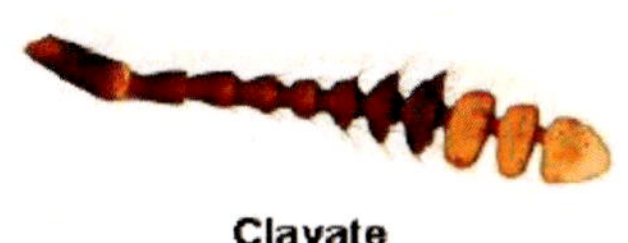

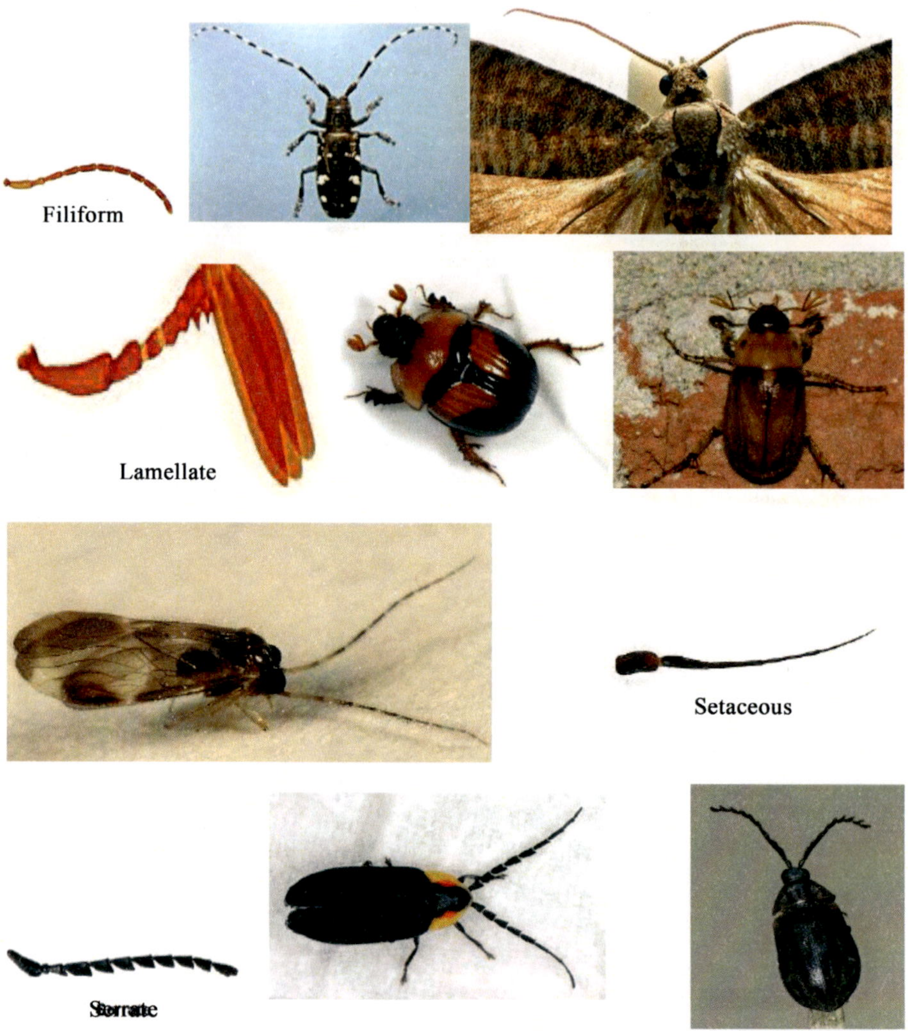
Filiform
Lamellate
Setaceous
Serrate

6

Photoreception

Light spectrum holds rich information about the world. The spectrum of natural illumination changes with daily cycle and objects differ in spectral reflection or emission properties. Like many other animals, insects use chromatic information to find favorable habitat, to efficiently locate food sources and to identify conspecific mates using the spectral information. Pollinating insects such as many species of bees and butterflies use their color vision to maximize success in foraging. They detect flowers, memorize the colors and patterns of rewarding flowers, and preferentially collect nectars from the flowers in their later visits. Visual guidance of behaviour is challenging when photons, the elementary particles of light, are scarce. To produce a reliable representation of the surroundings, the visual system must (i) ensure absorption of a sufficient number of photons into a photoreceptor, (ii) house photoreceptors that efficiently convert each photon absorption into a neural signal, and (iii) process these signals appropriately. Insects are a numerous and diverse class of arthropods that have evolved to occupy ecological niches from the brightest to the darkest.

1. Photosensitive cells in body wall (dermal detection)
2. Dorsal Ocelli and lateral stemmata
3. Compound eyes

Photosensitive cells in body wall = *dermal detection* e.g. hawkmoths

Ocelli/Simple Eyes

- Present in nymphs and adults of **Hemimetabolous insects** and adults of **Holometabola.**
- Usually 3 in number arranged in triangular fashion between the compound eyes (frons and vertex) or Vertex of the head.
- Consists of a single cornea secreted by the corneagen cells, below which are a group of retinular cells forming the rhabdom. These ocelli function as stimulatory organs to improve the sensitivity of the compound eyes. Single Corneal lens
- Each Ocellus have a **Biconvex lens**

- One to many photoreceptors
- The dorsal ocelli are represented by **fenestrae** in cockroach.
- The ocelli are much more sensitive than compound eyes. They measure light intensity and the information derived from them may be used to modify the insect's response to stimuli received by the compound eyes. Dorsal ocelli perceive light to maintain diurnal rhythm and is not involved in image perception.

Lateral ocelli (Stemmata)

- **Present only in endopterygote insect larvae** and may occur singly or in groups on either side of the head. They vary from 1-6 in number and some times 7 on each side. Lateral ocelli consist of cornea, a crystalline cone body and retinular cells forming the rhabdom. They differ from the dorsal ocelli in the fact that thay are innervated from the optic lobes of the brain. It helps to detect form, colour, distance and movement, and also to scan the environment.

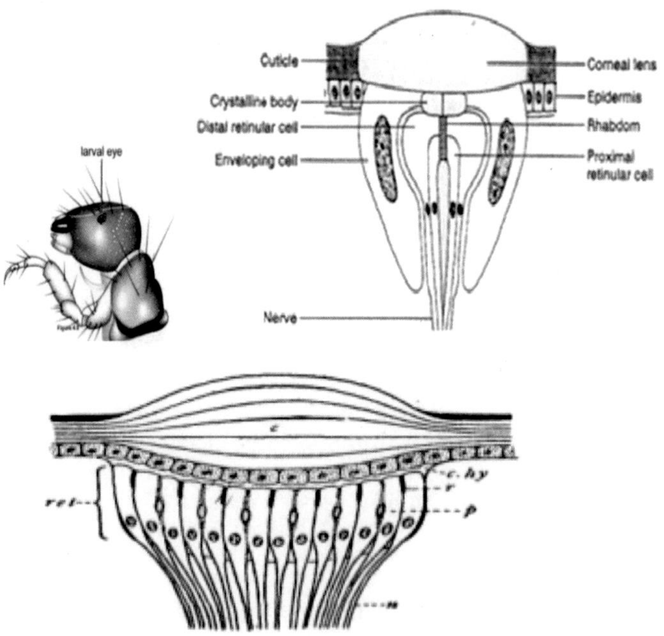

Simple eyes

Dorsal ocelli	Lateral ocelli
Found in adult insects and the larvae of hemimetabolous insects	Visual organs of larval holometabolous insects
Three ocelli forming are inverted triangle antero-dorsally on the head	It is laterally on the head and vary in number from one in sawfly larvae to six on each side in lepidopteran larva
The ocelli are lost or absent in wingless form	Stemmata are of two types: a. Those with single rhabdom: eg: mecoptera, neuroptera, etc. b. Those with multiple rhabdom: eg: grubs of adephaga suborder of beetles, sawfly larva,etc.
Ocelli are adapted for the concentration of light and perception of changes in intensity, a pathway for rapid conduction.	

Compound Eyes

These organs possess the ability to perceive light energy and able to produce a nerve impulse. The compound eyes may be completely absent in insects like **Protura** or they may remain reduced in endoparasitic **Hymenoptera, Siphunculata, Siphonaptera, female coccids** etc. The compound eyes are present on either side of the head capsule of an adult insect and also in the nymphs of Exopterygota. These are a pair and consists of number of individual units (or) facets called **ommatidia**. The number of ommatidia varies from 1 in the worker of ant, *Ponera Punctatissim a* to over 10,000 in the eyes of **dragonflies.** Each ommatidium is marked externally by a hexagonal area called facet. The shape of compound eye vary based on number of ommatidia. If the number of ommatidia is more they remain closely packed and they attain a **hexagonal** shape. If they are few, they remain loosely packed and they attain **circular** shape.

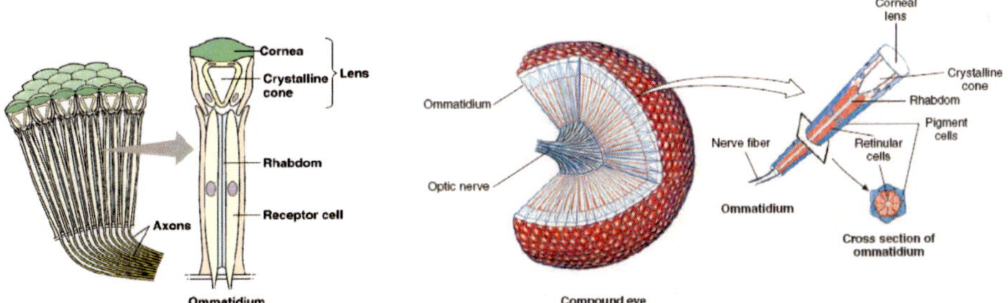

Structure of ommatidium

Ommatidium consists of 2 parts.

- Optic/Dioptic apparatus : Acts as lens
- Sensory/Receptor apparatus : Forms the image
- Each ommitidium consists of outer light gathering (optical) part (Cornea) comprising a **Biconvex lens** or cornea and a transparent cone underneath secreted by the surrounding **semper's cells**.

Optic/dioptic apparatus

• **Cornea** : It is a cuticular transparent colourless layer that remains continuous with the integument. The cornea comprised a **Biconvex lens** or cornea and a transparent cone underneath secreted by the surrounding **semper's cells.**

• **Corneagen cells**

- These are the modified epidermal cells which secrete the cornea and are two in number.

• **Crystalline cone cells**

- These cells remain just beneath the cornea and corneagen cells and are four in number, forming the crystalline cone and consists of a translucent material.

• **Primary pigment cells (or) Iris pigment cells**

- These are darkly pigmented cells, commonly two in number, present around the crystalline cone which are mainly useful for separating the ommatidia from one another and also restrict the movement of light passing from the neighboring ommatidia.

Sensory/receptor apparatus

• **Retinular cells**

- These are 6-10 in number (commonly 8) which are arranged and contribute to the formation of a centrally located rod like **rhabdom** (with the rhabdomeres (microtubules) which are formed with the inner side margins of retinular cells) on which the image is formed. The rhabdom contains a light absorbing pigment called **rhodopsin**. The retinular cells continue with the axons that pass through the basement membrane forming an optic nerve which remain connected to the optic lobes of the brain.

- **Secondary pigment cells**
  - The rhabdom is surrounded by secondary pigment cells that help to separate the ommatidia. They surround the retinular and primary pigment cells. These are numerous in number. Each ommatidium is covered by a ring of light absorbing pigmented cells, which isolates an ommatidium from other. Nerve cells are clustered around the longitudinal axis of each ommatidium. Each plugged into **optic lobe (of protocerebrum)**, thus each facet is thought to produce a distinct image.
  - The **resolving power** in insects is $\mathbf{1^0}$.
  - **Mosaic theory of insect vision** was proposed by **Muller (1829)**

Types of ommitidium

a. **Eucone type :** Semper cells produce hard crystalline cone. E.g. Lepidoptera

b. **Pseudocone :** Semper cells produce liquid filled extracellular cone. E.g. Diptera

c. **Acone :** Semper cells do not produce crystalline cone but have a clear cytoplasm which occupies the region of the cornea. Eg.Hemiptera

d. **Exocone :** Semper cells do not produce a crystalline cone but extends inwards to the retinular cells as slender retractile strands. E.g. Coleoptera

Types of Insect Vision

Apposition eyes / Photopic eyes/Day eyes	**Superposition eyes / Scotopic eyes/ Night eyes**
Rhabdom extends the full length of ommitidium from crystalline cone to basement memberane and the distribution of pigment in the pigment cells is little affected by the condition of illumination	Rhabdom is connected to crystalline cone by a transluscent filament and is confined the basal half of the ommitidium
Ommitidia forms image side by side as a mosaic with rhabdom extending full length (**Distinct image**)	Overlapping of images (**Dim image**)
Found in diurnal insects	Found in **nocturnal and crepuscular insects**

Binocular vision : Both Apposition and Superposition images formed (**Preying mantis**)

Color Vision in Insects

- Not all insects see colors
- Insects don't see across entire color spectrum like vertebrates

- Bees only can be trained to discriminate between about four colors
- **Honeybees** can differentiate yellow, blue green, blue violet, ultraviolet and bees purple but are **red colour blind**.
- Many insects don't see yellows well (this is how bug lights work)
- **Butterflies** have among the **broadest color sensitivity** of any animal, i.e., 300 to 700 nm

Detection of polarized light

- Sunlight get increasing polarized at greater angles from the sun, i.e., the more the light vibrates in some planes relative to others
- Light most polarized at 90 degree angle to sun
- Insects can detect polarized (UV) light (different portions of sky differently colored)
- A single patch of sky is enough for some insect to calculate the sun's position at any time of day
- Polarized light important in insect navigation, e.g., ants and social insects

Image Formation

- Insects are sensitive to movement but may have difficulty detecting still objects
- With regard to movement; bugs can distinguish single degree of arc
- Insects (e.g., bees) learn shapes and colors

7

Thorax

- The thorax is the organ of locomotion due to articulation of wings and legs and is connected with the by neck or cervix.
- Thoracic segments are made up of three sclerites namely, dorsal body plate **tergum or nota**, ventral body plate **sternum** and lateral plate **pleuron**.
- Second and middle tagma which is three segmented, namely prothorax, mesothorax and metathorax. Meso and metathorax which bear wings are called as **Pterothorax**.
- In the apterygote insects the three thoracic segments are similar in size and structure. In the pterygote insects the thorax is highly modified.
- The **prothorax** is quite large and is somewhat independent of the other segments. The large shield that covers the dorsal and lateral sides of the prothorax is the **pronotum**.
- The pronotum is the dorsal sclerite of the prothorax, which can be highly modified in various orders such as the Hemiptera, Blattaria, and Coleoptera. E.g undivided and **Saddle** shaped in grass hopper, **Shield** like in cockroach.
- *Cockroaches* have *pronotums that extend forward over the head.*
- Scarab beetles and other beetles may also have unusual pronotums.
- Treehoppers have some of the most bizarre pronotums of all insects.

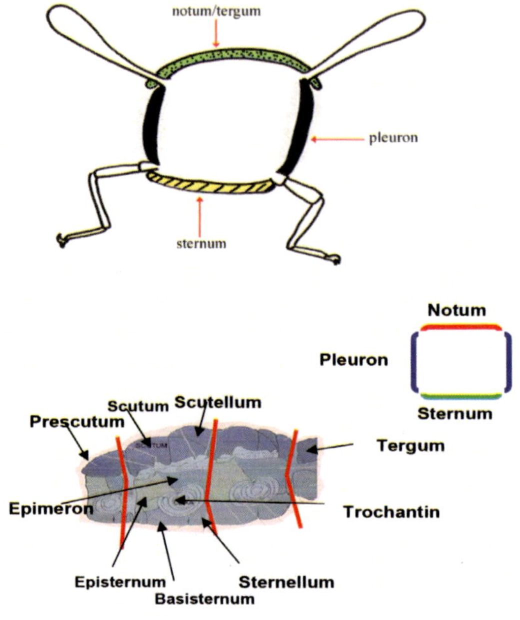

Pronotum modifications

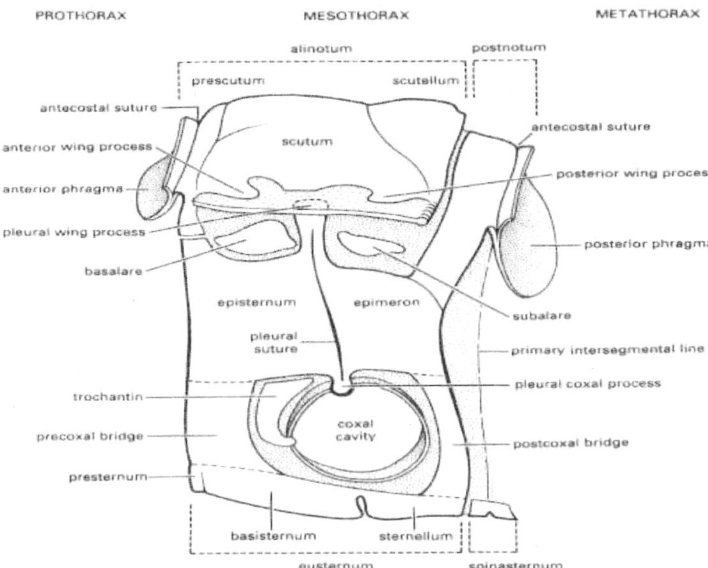

Diagrammatic lateral view of a wing-bearing thoracic segment

- ## Pterothoracic notum
  - Have 3 transverse sutures (antecostal, prescutal and scuto-scutellar) and 5 tergites (acrotergite, prescutum, scutum, scutellum and post-scutellum).
  - The **prescutum** is the narrow anterior sclerite with expanded lateral areas.
  - The **scutum** is posterior to the prescutum and is typically the largest and most prominent tergite.
  - The **scutellum** is the posterior sclerite that is formed by a V-shaped suture. Together these three sclerites (prescutum, scutum and scutellum) form the **alinotum, the wing bearing portion of the notum**.
  - Posterior to the alinotum is the **postnotum**.
  - Internally the postnotum bears a large plate **(phragma)** that provides a large surface area for the attachments of the dorsal longitudinal muscles.
- **Thoracic pleura**:
  - Lateral body wall of thoracic segment between notum and sternum.
  - Selerites of pleuron is called as pleurite and they fuse to form pleural plate. Pleural plate is divided into anterior **episternum** and posterior **epimeron** by **pleural suture**.

    - Pterothoracic pleuron provides space for articulation of wings and legs.

- **Thoracic sterna**
    - Ventral body plate of each thoracic segments are called as prosternum, mesosternum and metasternum.
    - Thoracic sterna is made up of a segmental plate called **eusternum** and an intersternite called **spinasternum**.
    - Eusternum is made up of three sternites viz., presternum, basisternum and sternellum.
    - Two pits connected by a suture mark the posterior edge of the basisternite. These **furcal pits** mark an invagination of the cuticle forming the **furcal arms** (=sternal apophyses) an internal skeletal rod.

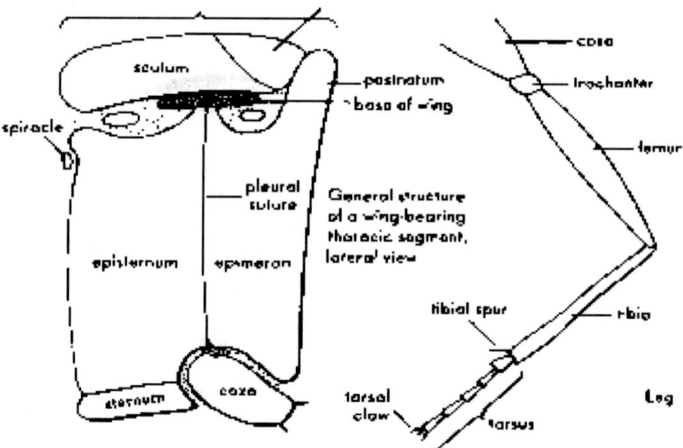

Thorax and leg structure.

Thoracic Appendages

Legs

A pair of legs is attached to each segment. The first pair of legs attached to prothorax is called as prothoracic legs or forelegs. Second pair of legs attached to mesothorax is called as mesothoracic legs. And the third pair of legs attached to metathorax is called as metathoracic legs or hind legs.

Wings

- Two pairs of wings are present in true insects.
- One pair attached to mesothorax is called as mesothoracic wings or fore wings and pair of wings attached to metathorax is called as metathoracic wings or hind wings.

- In some insects like house fly, only one pair of wings is present. It is either attached to mesothorax or metathorax. When this pair is attached to mesothorax, mesothorax is well developed than metathorax and vice-versa.
- Two pairs of spiracles are also present in the mesopleuron and metapleuron.

Insect Legs

- The **fore-legs** are located on the prothorax, the **mid-legs** on the mesothorax, and **the hind legs** on the metathorax. Each leg has six major components, listed here from proximal to distal: **coxa** (plural coxae), **trochanter**, **femur** (plural femora), **tibia** (plural tibiae), **tarsus** (plural tarsi), **pretarsus**.
- The femur and tibia may be modified with spines. The tarsus appears to be divided into one to five "pseudosegments" called **tarsomeres**. Like the mouthparts and antennae, insect legs are highly modified for different functions, depending on the environment and lifestyle of an insect.
- In the leg of an insect, the cuticle may contain unicellular hair–like outgrowth known as Setae, which represent sensory organs of various types and may also take the form of pegs, hooks or scales.
- The femur and tibia may be modified with spines (multi-cellular outgrowths).
- Multi-cellular outgrowths are of two types. If they are immovable they are called spines, if they are movable and articulated they are referred as spurs.
- Definite impressed lines or internal ridges between sclerites are known as sutures. Flexible in folding of the body wall are called conjunctivae and definite joints as between segments of legs are called articulation.
- The flexible portion of the cuticle, connecting the hard ring like portions of any two segments is called conjuctive and articular membrane.

A typical leg of an insect consists of the following parts

- **Coxa** : (Pl. coxae)
    - It is the first or proximal leg segment. It articulates with the cup like depression on the thoracic pleuron. It is generally freely movable.
- **Trochanter**
    - It is the second leg segment. It is usually small and single segmented. Trochanter is two segment of the dragonfly, damselfly and ichneumonid wasp (parasitic hymenoptera). The second trochanter is called **trochantellus.**

- **Femur** : (Pl. femora)
    - It is the **largest and stoutest/strongest** part of the leg and is closely attached to the trochanter. It contains the main muscles used in running, jumping and digging.
- **Tibia** : (Pl. tibiae)
    - It is usually long and provided with downward projecting spines which aid in climbing and footing. Tibia of many insects is armed with large movable spur near the apex.
- **Tarsus** : (Pl. tarsi)
    - Foot of the insect leg and can consist of between one and five segments.. The sub segment of the tarsus is called tarsomere. The basal (1st) tarsal segment is often large, big or broader than others and is named as **basitarsus**.
    - In some insects it bears 2, or 3 or 4 or one segment.
    - Tarsus with single joint – Ex. Human louse
    - Tarsus with two joints – Ex. Aphid
    - Tarsus with three joints – Ex. Mole cricket, Gryllidae (grass hopper)
    - Tarsus with four joints – Ex. Tettigonidae (grasshopper), Leaf beetle.
- **Claws**
    - Beyond the tarsus there are several structure collectively known as pretarsus. Tarsus terminates in a pair of strongly curved claws with one or two pads of cushions at their base between them.
    - A pad known as an 'arolium' present between the claws help the insect to hold the smooth substrates.
    - The tarsus usually consists of several small joints, the last of which generally carries a pair of terminal **claws**. The last tarsal segment may be extended between the claws to form a pad-like organ - the **Arolium** (Grasshopper).
    - Some Diptera (true flies) have two additional pads - the **Pulvilli** (singular: **pulvillus**) - lying below the claws on either side of the arolium, although in most flies, including the common housefly and its relatives, the arolium is replaced by a stout central bristle - the **empodium**. In many insects, the other tarsal segments also have ventral pulvillus-like organs, called **plantulae**.
    - These structures produce a sticky secretion and act like 'suction-pads', enabling the insect to climb smooth or steep surfaces.

- In some insects, the ventral surface of pretarsus consist of a median circular plate between the claws known as **unguitractor** where as the claws are known as **ungues**.

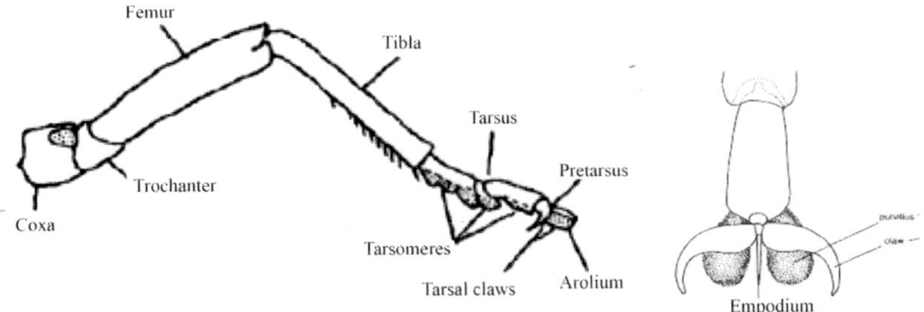

Legs modifications

Like the mouthparts and antennae, insect legs are highly modified for different functions, depending on the environment and lifestyle of an insect.

Type of leg	Remarks	Example
Ambulatorial (Walking leg)	• Legs are used for walking. • Coxa widely separated • Femur and tibia are long	Bugs (order Hemiptera), leaf beetles (order Coleoptera).
Cursorial (Running leg)	• legs are modified for running. • All the legs uniformly well developed without any special modification • Femur is not swollen at a speed of a cockroach moves 4.6 km/hour	Cockroaches (order Blattaria), ground and tiger beetles (order Coleoptera).
Fossorial (Digging or Burrowing leg)	• Fore legs are modified for digging. • Femur is stout. • Tibia is short and stout and bears distally two or three strongly pointed tines. • The first two segments of tarsus are also produced into strong tines. The tine of basistarsus work against one of the tibial tines to function as shears in cutting fine rootlets. • Tympanum is present in fore tibia.	Ground dwelling insects; mole crickets (order Orthoptera) and cicada nymphs (order Hemiptera).
Natatorial (Swimming leg)	• Legs are modified for swimming. These legs have long setae on the tarsi. • These legs have long setae on the tarsi. • Hind legs are modified. • Tibia and tarsus short and broad and are provided with dense long marginal hairs. • Coxa flattens out on the body wall. Tibia and tarsus are flat.	Aquatic beetles (order Coleoptera) and bugs (order Hemiptera).

Type of leg	Remarks	Example
Raptorial (predatory or Grasping leg)	• Forelegs modified for grasping (catching prey). • Coxa is long and mobile. The elongated coxae extends to capture the prey. • The femur is large and grooved along the ventral surface with spines on the two opposing surfaces of femur. • The spiny tibia fits into the femoral groove when it snaps down over the prey	Mantids (order Mantodea), ambush bugs, giant water bugs and water scorpions (order Hemiptera).
Saltatorial (Leaping or Jumping Leg)	hind legs adapted for jumping. These legs are characterized by an elongated femur and tibia.	Grasshoppers, crickets and katydids (order Orthoptera).
Foragial (pollen collecting)	• Tibia of hind leg is modified for pollen collecting and consists of Pollen basket, Pollen packer and Pollen comb. • Pollen basket (corbicula) enables the bee to carry a larger load of pollen and propolis from the field to the hive. • Pollen packer (pollen press) is a row of stout bristles at the distal end of tibia. Auricle is a small plate fringed with hairs at the. Pollen packer is useful to load pollen in corbicula. • Pollen comb is used to collect pollen from middle legs and from posterior part of the body	Honeybee (Hymenoptera)
Antennal cleaning First pair legs of honey bee	• Tibia possesses a process and the 1st tarsal segment of fore-legs possesses a semi-circular notch. • Each thoracic leg has a row of stiff bristle on tibia forming an eye brush for • cleaning the compound eyes. At the distal end of tibia there is a movable spine the velum which can close over a notch on the tarsus, to form an antenna comb, through which the antenna is drawn for cleaning.	Honeybee (Hymenoptera)
Wax picking type Second pair of legs in Honey bees	• *Long bristles on the meso-thoracic leg tarsus form a pollen brush* for removing pollen from the front part of the body. • Each mesothoracic leg has a pollen brush on the tarsus; the end of the tibia has a spur like spine for removing pollen from the pollen basket and wax from abdomen. Hence the *tibial spine is called wax pick.*	Honeybee (Hymenoptera)

Type of leg	Remarks	Example
Scansorial (Clinging type legs)	• Legs are modified for clinging to the hairs of its host. • Tibia is stout and at one side bears a thumb like process. • The tarsus is single segmented. • There is a single large claw that usually fits against a thumb like process which forms an efficient mechanism for hanging to the hairs of host	Head louse
Sticking Leg	• Pretarsus consists of a pair of lateral adhesive pads under the claws called pulvilli. • Arolium is absent. But a median spine like structure empodium is present. • The pulvilli arecovered with dense mats of tiny glandular hairs called tenant hairs. • Secretions of these glandular hairs are helpful in clinging to smooth surface and to walk upside down on the ceiling.	House fly (Diptera)
Clasping Leg	• The tarsus is flattend with adhesive discs which are useful to clasp the mate during copulation.	male water beetle
Prehensile (Catching Prey/ basket forming type)	• All legs are modified • Thoracic segments obliquely arranged. Tergal plates are pulled backwards and Sternal plates pushed forward, resulting that all the legs pushed forward and seen below the head, together from a basket like structure useful for catching the prey even in flight	Dragonflies

Cursorial Raptorial Fossorial

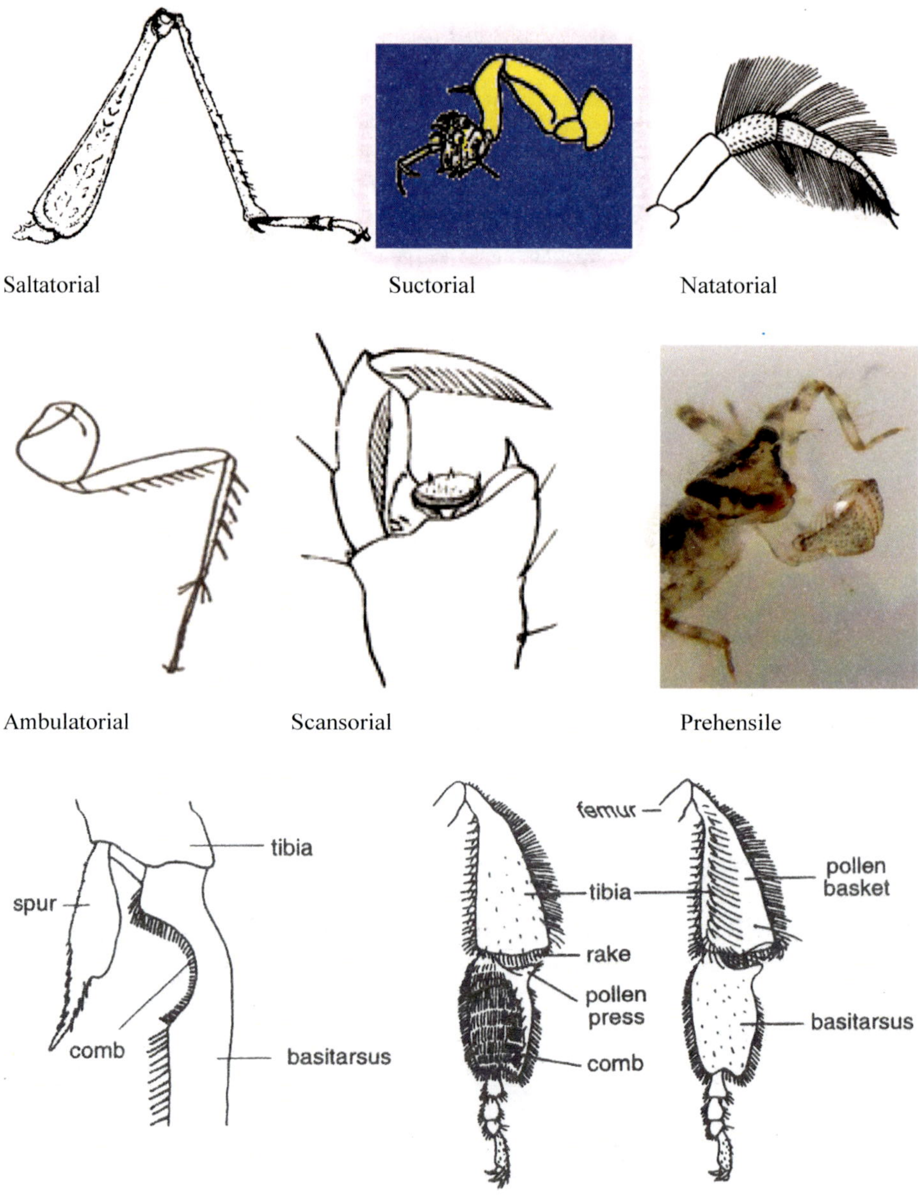

Saltatorial Suctorial Natatorial

Ambulatorial Scansorial Prehensile

Antennal cleaning (1st segment) Foragial (Tibia of hind leg)

Legs of immature stages

The immature stage of **exopterygotes** i.e. nymph consist of only thoracic legs similar to its adult where as that of **endopterygote** i.e. larva possess two types of legs.

- Thoracic legs or **true legs**
  - Jointed, present on all the 3 thoracic segments.
- Abdominal legs or **prolegs**
  - Unjointed sucker like legs, having flat, fleshy surface at its tip known as **planta**. The planta consists of hook like structures known as **crochets** which are used for clinging to the substrate.
  - The number of prolegs vary from 1-5 pairs which are distributed on 3rd, 4th, 5th, 6th and 10th abdominal segments.
  - Two to five pairs are normally present. They are unsegmented, thick and fleshy.
  - **Caterpillars**: Larvae having prolegs on 3,4,5,6 and 10th abdominal segment
  - **Semilooper** : Larvae having prolegs on 5, 6 and 10th abdominal segment
  - **Looper** : Larvae having prolegs on 6 and 10th abdominal segment
  - **Sawfly** (Tenthridinidae, Hymenoptera) : Larva have 8 pair of prolegs. This is the unique feature of sawfly larva, but these prolegs do not bear crochets, unlike lepidopteran larva.

In some insects leg are degenerated e.g.: Coccidae; Endoparasitic hymenopterans.

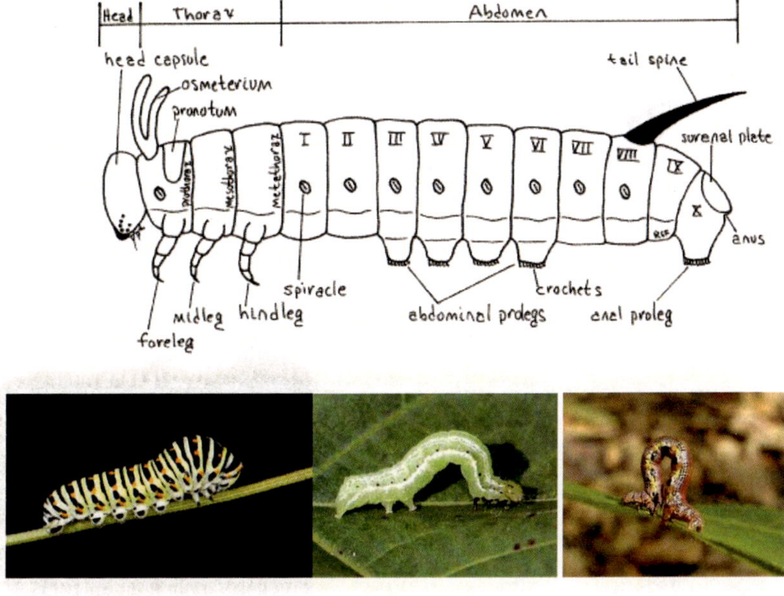

Caterpillar Semilooper Looper

Insect Wings

- **Wings** are a flattened double - layered expansion of body wall with a dorsal and ventral lamina having the same structure as the integument.
- The success of the insects as terrestrial animals is atleast partly due to their ability to fly.
- Wings, their modifications and the characters of wings are the primary tools for insect taxonomic studies, and for the identification of insects.
- Typically, adult insects have two pairs of wings articulating with the thorax and consisting of flattended lobes of the integument supported by hallow veins.
- Most insects have two pairs of wings-one pair on the mesothorax and one pair on the metathorax (never on the prothorax). The wing-bearing thoracic segments (meso-& meta-thorax) together called as pterothorax.
- Silver fish and spring tails do not have wings (Apterygota; **Primarily Wingless**)
- Some insects have only one pair of functional wings (Diptera)
- Wings are deciduous in ants and termites. The insects break off or tear off their wings after a single nuptial flight and before beginning their life in soil. (**Secondarily wingless**)
- Fleas, lice and certain aphids and ants have degenerate wings
- Insects are the only invertebrates that can fly.
- Their wings develop as evaginations of the exoskeleton during morphogenesis but they become fully functional only during the adult stage of an insect's life cycle.
- The wings may be membranous, parchment-like, heavily sclerotized, fringed with long hairs, or covered with scales.
- The base of the wing connects with the body by a membranous hinge and membranous hinge bears a group of small sclerites called axillary sclerites. Axillary sclerites articulate with the edge of the notum.
- Wings serve not only as organs of flight, but also may be adapted variously as protective covers (Coleoptera and Dermaptera), thermal collectors (Lepidoptera), gyroscopic stabilizers (Diptera), sound producers (Orthoptera), or visual cues for species recognition and sexual contact (Lepidoptera).
- Fully developed and functional wings occur only in adult insects, although the developing wings may be present in the larvae. In hemi-metabolous

nymphs, they are visible as external pads (exo-pterygota), but they develop internally in holometabolous forms (endo-pterygota). The Epimeroptera are exceptional in having two fully winged nymphal instars.

- In most cases, a characteristic network of veins (wing venation) runs throughout the wing tissue.
- These veins are extensions of the body's circulatory system. They are filled with hemolymph and contain a tracheal tube and a nerve.
- In membranous wings, the veins provide strength and reinforcement during flight.
- Wing venation is a commonly used taxonomic character, especially at the family and species level. Wing shape, texture, and venation are quite distinctive among the insect taxa and therefore highly useful as aides for identification.
- Two orders of winged insects, the Ephemeroptera and Odonata, have not evolved wing-flexing mechanism (folding wings over body during rest), and their axillary sclerites are arranged in a pattern different from that of the Neoptera; these two orders (together with a number of extinct orders) form the Paleoptera.

Wings Development

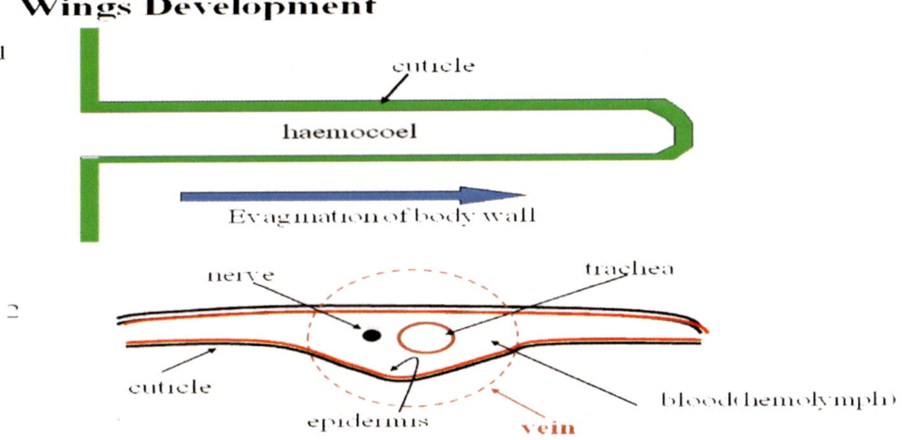

- Both dorsal and ventral laminane grow, meet and fuse except along certain lines to form a series of channels which serve for the passage of tracheae, nerves and blood.
- The walls of these channels become thickened to form veins or nervures. Wing is nourished by blood circulating through veins.
- The arrangement of veins on the wings is called **venation**.

- In insects like dragonfly and damselfly, there is an opaque spot near the coastal margin of the wing called **pterostigma**.
- Wings tend to have less venation and be more stiff in is insects with faster wing speed
- Present on meso and meta thoracic segments. Apodemes serve as interior attachment sites for muscles
- Insects have evolved many variations of the wings, and an individual insect may possess more than one type of wing.
- Wing venation is a commonly used taxonomic character, especially at the family and species level.
- Not all insects have wings 5 primitive orders have never developed wings (known as Apterygote insects) ex. collembolans, springtails highly specialized insects without wings parasites such as fleas and lice castes of social insects--worker ants, termites functional wings occur only in the imago

Several hypotheses have been proposed for the evolution of wings

- In one hypothesis, wings first evolved as extensions of the cuticle that helped the insect absorb heat and were later modified for flight.
- A second hypothesis argues that wings allowed animals to glide from vegetation to the ground.
- Alternatively, wings may have served as gills in aquatic insects.
- Still another hypothesis proposes that insect wings functioned for swimming before they functioned for flight.

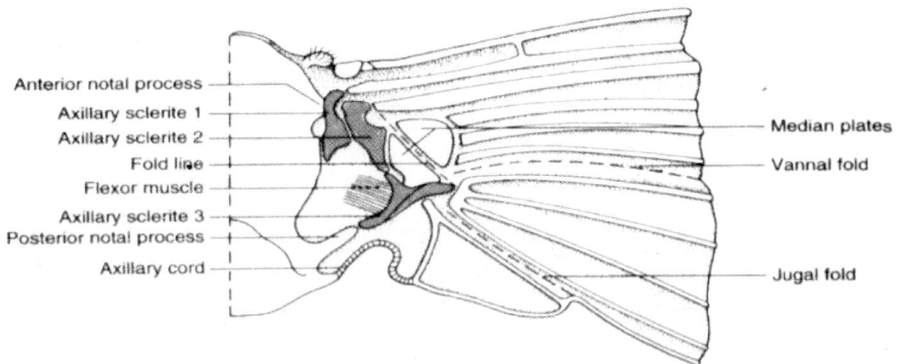

Wings margins and angles

- The wing is triangular in shape and has therefore three sides and three angles.
- The anterior margin strengthened by the costal is called coastal margin and the lateral margin is called apical margin and the posterior margin is called anal margin.
- The angle by which the wing is attached to the thorax is called humeral angle.
- The angle between the coastal and apical margins is called apical angle.
- The angle between apical and anal margins is anal angle.

Wing regions

- The anterior area of the wing supported by veins is usually called **remigium**.
- The flexible posterior area is termed **vannus**.
- The two regions are separated by **vannal fold**.
- The proximal part of vannus is called **jugum**, when well developed is separated by a **jugal fold**.
- The area containing wing articulation sclerites, pteralia is called axilla.

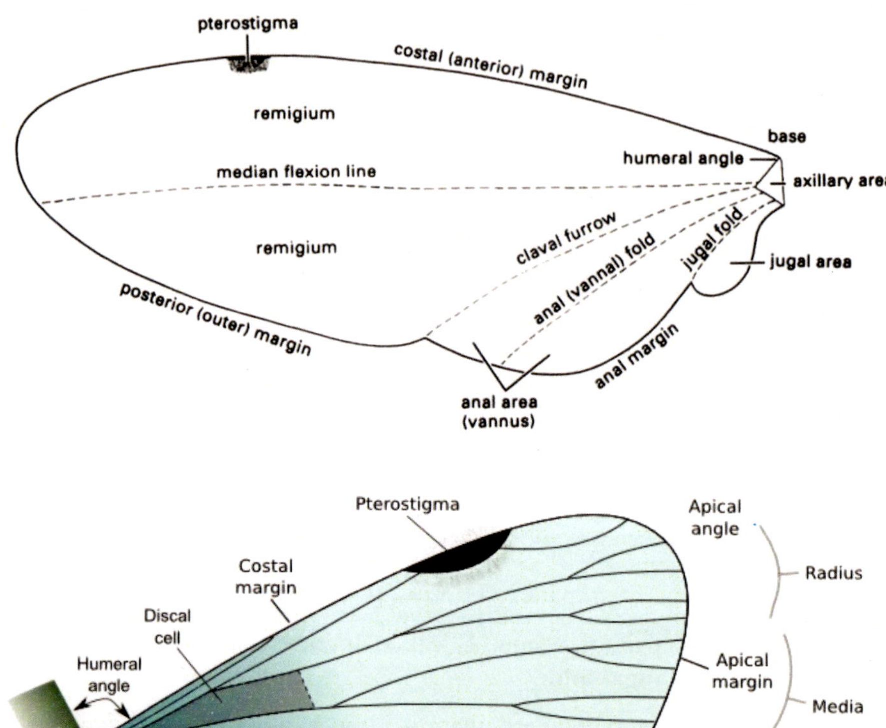

Wing Venation

- The **archedictyon** is the hypothetical scheme of wing venation based on a combination of speculation and fossil data.
- The *archedictyon* is the name given to a hypothetical scheme of wing venation proposed for the very first winged insect such as fossil insects. Since all winged insects are believed to have evolved from a common ancestor, the *archedictyon* represents the "template" that has been modified (and streamlined) by natural selection for 200 million years. According to current dogma, the archedictyon contained 6-8 longitudinal veins. These veins (and their branches) are named according to a system devised by *John Comstock and George Needham*-the Comstock-Needham System.

- The *Comstock-Needham system* is a naming system for Insect Wing Veins, devised by *John Comstock* and *George Needham* in *1898.* It was an important step in showing the *homology* of all *insect wings.* This system was based on Needham's *pretracheation theory* that was later discredited by *Frederic Charles Fraser* in 1938.
- Insect wings have rigid veins which support the wing in flight. The wing veins may look different in different insect groups, scientists tracked that all different insect wings are evolved from the same ancestor, i.e. wings had evolved only once in the insects history.

Typical insect wing venation (modified Comstock-Needham System)

- The principal longitudinal veins arranged in order from the anterior margin are costa (C), sub costa (Sc), radius (R), median (M), cubitus (Cu) and anal veins (A).
- Small veins often found inter connecting the longitudinal veins are called cross veins.
- Due to the presence of longitudinal veins and cross veins, the wing surface gets divided into a number of enclosed spaces termed **cells**. Cell is the areas of various shapes enclosed between the veins are called "*Cells*". Cells are of two types: *Open cells and closed cells*. If the area is entirely surrounded by veins–"closed cells" and if the area extends to the wing margin without intervening veins–"Open cell".
- Generally the veins are heavier or closely placed towards the costal margin, due to greatest stress during flight.
- **Precoasta (PC)** -- This vein is fused with costa in all extant insects, mostly unrecognisable.
- **Costa (C)** -- at the leading edge of the wing, strong and marginal, extends to the apex of the wing, it is unbranched.
- **Subcosta (Sc)** -- the second longitudinal vein, mainly the subcosta posterior sector **(ScP)**. **Sc** is reduced or fused with **R** in most Hemiptera.
- **Radius (R)** -- the third vein, usually the strongest vein on the wing, with branches usually cover the largest area of wing apex. **RP** is often referred to as radial sector **(Rs)** and the end branches as **R1-5**.
- **Media (M)** -- the fourth longitudinal vein, **MA** and **MP** usually with 4 branches each. In some insect groups **MA** fused with **R** so only **MP** on the medial area. In this case the **MP1-4** are often referred as **M1-4**.
- **Cubitus (Cu)** -- fifth longitudinal vein, CuA may branch to 4 or fewer veins.

CuP is unbranched, lies near the claval fold and reach the wing posterior margin.

- **Anal veins (A)** -- veins behind the cubitus, **AA** and **AP** are usually separated by the anal fold. In Neoptera, **AA** is always fused with **Cu** or **CuP**. In the hind wings of most orthopteroid insects, there is a large anal area where anals branch several times to form a fan-like folded wing.
- **Jugal (J)** -- small veins in the jugal area, found only in Neoptera.
- The black **pterostigma** is carried near the wing tip, between RA1+2 and RA3+4.

Cross-veins

- **Cross-veins** are transverse veins joining longitudinal veins. Their names are based on the position relative to longitudinal veins, e.g. r-m is the cross-vein between radius and media.
- Names of cross veins are based on their position relative to longitudinal veins:
    - **c-sc** cross veins run between the costa and subcosta
    - **r** cross veins run between adjacent branches of the radius
    - **r-m** cross veins run between the radius and media
    - **m-cu** cross veins run between the media and cubitus

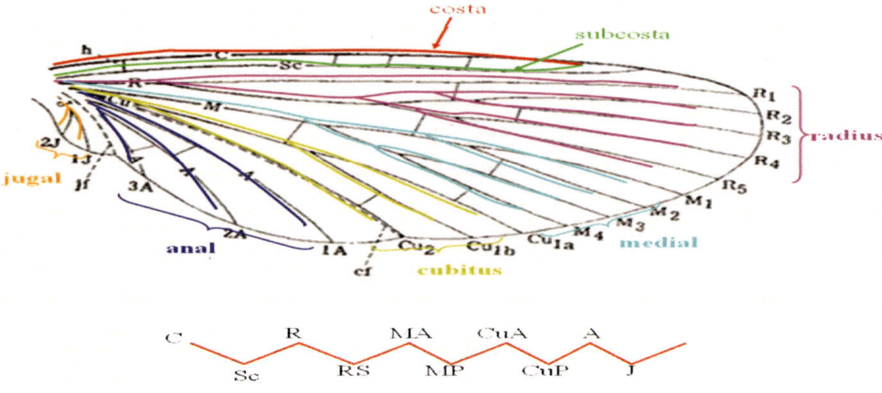

Evolution of Wings

Paranotal theory

- Wings developed from paranotal lobes that were first used for gliding.

Branchial theory

- Wings developed from gills that were subsequently used for swimming, and finally for flight.

Paleoptera	Neoptera
• Primitive insects • Unable to fold their wings at rest. • Insects carry their wings vertically or horizontally to their bodies. • axillary sclerites are arranged in a pattern different from that of the Neoptera • Ephemeroptera and Odonata	• Most modern insects • Insects are able to fold their wings over the body at rest. • This enables the insects to fit into smaller spaces. • three axillary sclerites that articulate with various parts of the wing. , a muscle on the third axillary causes it to pivot about the posterior notal wing process and thereby to fold the wing over the back of the insect • Lepidoptera, coleoptera, hemiptera etc.

A. Paleoptera **B.** Neoptera

Wing modifications

Wings types	Characteristic	Order(s)
Elytra (singular elytron)	• Fore wings are hard and sclerotized that serve as protective covers for membranous hind wings • Wing venation is lost. • During flight they are kept at an angle allowing free movement of hind wings	Coleoptera and Dermaptera
Hemelytra (Singular : Hemelytron)	• Fore wings are hardened throughout the proximal two-thirds, while the distal portion is membranous. • Unlike elytra, hemelytra function primarily as flight wings.	Hemiptera: Heteroptera (Bugs)
Tegmina (singular tegmen)	• Fore wings that are completely leathery or parchment-like in texture. • Like the elytra on beetles and the hemelytra on bugs, the tegmina help protect the delicate hind wings and are not used for flight.	Grasshoppers, crickets and katydids (order Orthoptera), Cockroaches (order Blattaria), Mantids (order Mantodea).
Halteres (Singular : Haltere)	• Small, club-like hind wings that serve as gyroscopic stabilizers during flight; used for balance and direction during flight. • Each haltere is a slender rod clubbed at the free end (capitellum) and enlarged at the base (scabellum). • On the basal part two large group of sensory bodies form the smaller hick's papillae and the large set of **scapel** plate.	Diptera (True flies)
Fringed wings	• Slender front and hind wings with long fringes of hair • Wings are usually reduced in size. • Wing margins are fringed with long setae.	Thysanoptera (Thrips)
Hairy wings	• Fore wings and hind wings clothed with setae	Trichoptera
Scaly wings	• The scales make the wings colourful. Scales are unicellular flattened outgrowth of body wall. • Scales are inclined to the wing surface and overlap each other to form a complete covering. • Scales are responsible for colour. • They are important in smoothing the air flow over wings and body	Butterflies, moths and skippers (order Lepidoptera), caddisflies (order Trichoptera).

Wings types	Characteristic	Order(s)
Membranous wings	• Wings are thin and more or less transparent, but some are darkened • They are thin and supported by a system of tubular veins. • In many insects either forewings or hind wings or both fore wings and hind wings are membranous. They are useful in flight.	Dragonfiles and damselflies (order Odonata), lacewings (order Neuroptera), flies (order Diptera), bees and wasps (order Hymenoptera), termites (order Isoptera).
Clefted Wings (Fissured Wings)	• *Front wing* is longitudinally divided forming a *fork-like* structure. • The *hind wing* is divided twice, forming two forks *with three arms.* • All forks possess small marginal hairs. • They are *useful for flight*	*Both wings of Plume Moth*

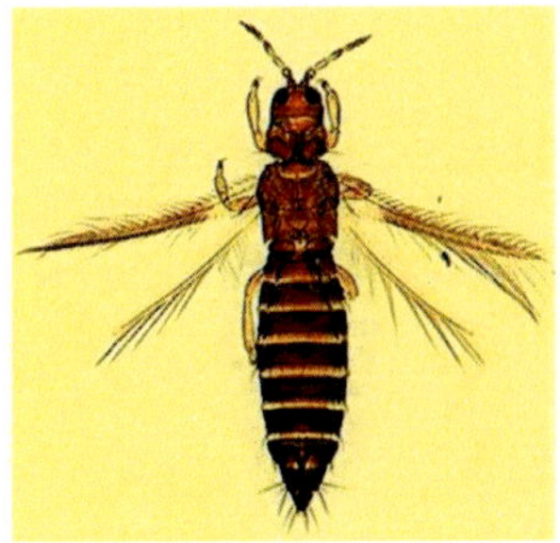

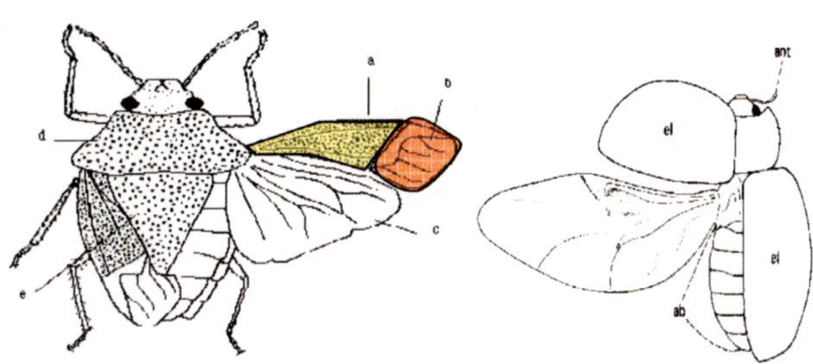

Fringed wings Hemelytra Elytra

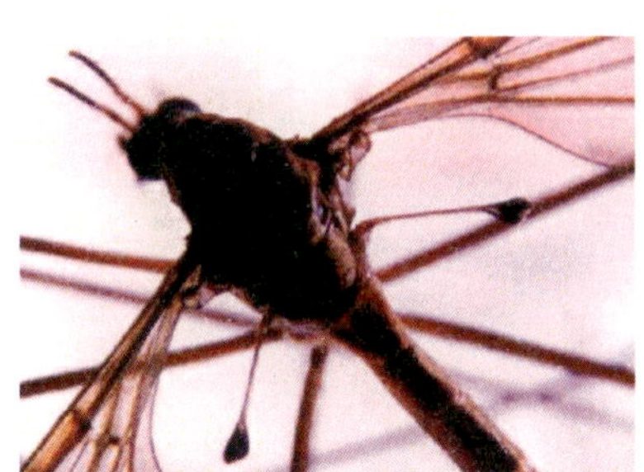

Halteres Tegmina

Wing coupling

- Higher pterygotes have devices for hooking fore and hind wings together so both pairs move synchronously. By coupling the wings the insects become functionally two winged.

- During flight, both the wings of insects are kept together by different inter-locking small structures. These inter-locking structures are called wing coupling apparatus.
- They are of different types: Jugum type, Frenulum & Retinaculum type, Amplexiform & Humuli type

Types of wing coupling

- **Hamulate**
  - A row of small hooks is present on the coastal margin of the hind wing which is known as hamuli.
  - These engage the folded posterior edge of fore wing.
  - e.g. bees

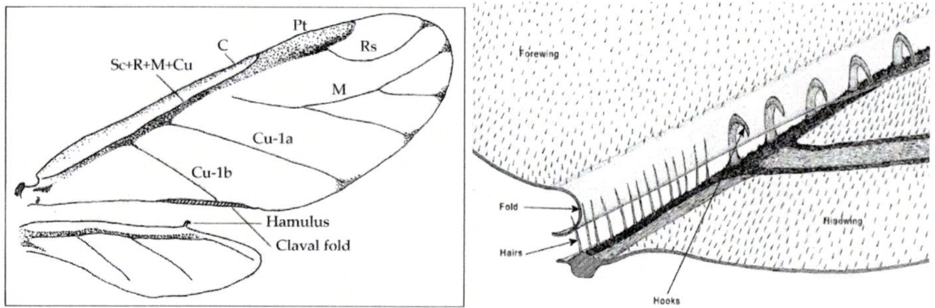

- **Amplexiform**
  - It is the simplest form of wing coupling.
  - A linking structure is absent.
  - Coupling is achieved by broad overlapping of adjacent margins.
  - e.g. butterflies.

Amplexiform coupling

- **Frenate** : There are two sub types. e.g. Fruit sucking moth.
- **Male frenate**
  - Hindwing bears near the base of the coastal margin a stout bristle called frenulum which is normally held by a curved process, retinaculum arising from the subcostal vein found on the surface of the forewing.
- **Female frenate**
  - Hindwing bears near the base of the costal margin a group of stout bristle (frenulum) which lies beneath extended forewing and engages there in a retinaculum formed by a patch of hairs near cubitus.

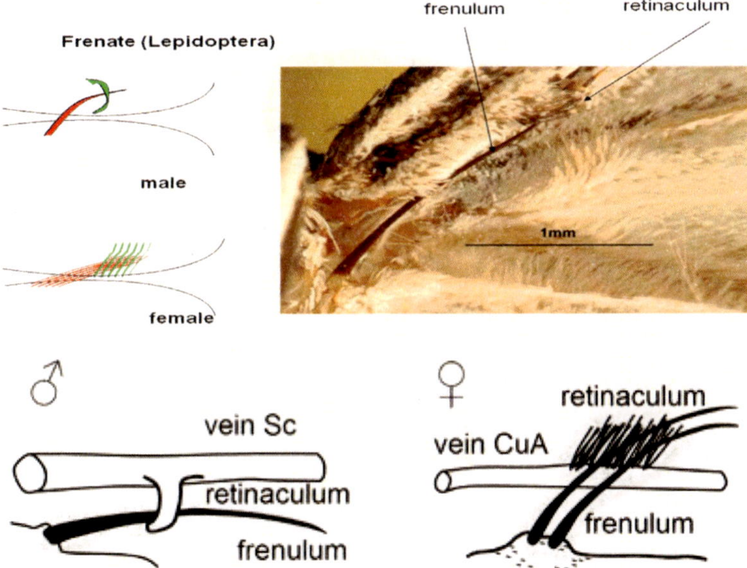

- **Jugate**
  - Jugam of the forewings are lobe like and it is locked to the coastal margin of the hindwings. e.g. Hepialid moths, *Primitive Lepidopterous Insects*.
  - The *anal margin of the forewing possesses a small lobe* (jugum lobe or jugum) at its base called fibula-, projects behind hind-wings, which rests upon the surface of the hind wing.
  - Sometimes a *spine is present on costal surface of hindwings*. In this manner the wings become inter locked.

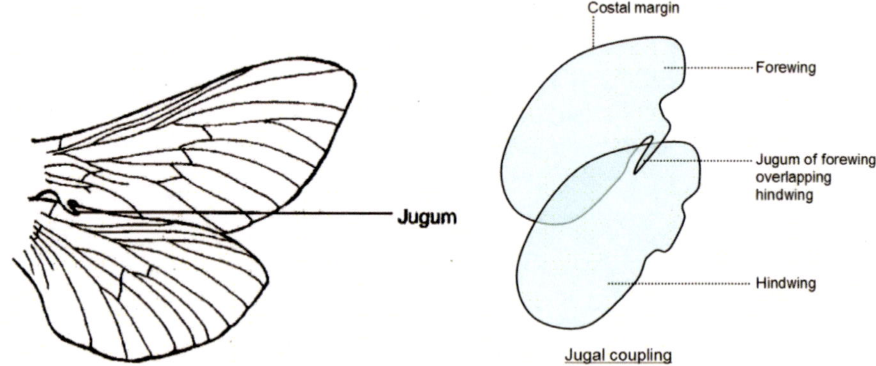

Jugum coupling

Flight Mechanism

Involves

- Flight muscles which aren't part of the wing
- Elastic property of a substance called resilin present in the hinges of wings
- Elastic properties of the thorax itself

Flight muscles: there are 2 types

- Direct : attached to the wing itself
- Indirect: not attached to the wing but to the notum (dorsal) and sternum (ventral)
- In all insects, the upstroke is affected by contraction of indirect flight muscles; this results in the notum being pulled down toward the sternum. A fulcrum point at the pleural process forces the wing up.
- In butterflies and moths, beetles, grasshoppers, cockroaches, dragon flies and others, the down stroke is powered by the direct flight muscles in conjunction with the relaxation of the indirect muscles
- In insects such as bees, wasps, and flies the down stroke is also achieved by indirect muscles called the dorsal longitudinal muscles

Insect wings movements

Insects	Beats/seconds	Flight (km/hr)
Dragonfly	20-28	26
Beetles	45-90	5
Butterflies	9-12	9
Hawkmoth	70	18
Mosquitoes	300-550	32
Midges	1000	22
Honeybees	200	22

Movement	Body lengths per second
Human	5
Volkswagen 'beetle'	5
jet fighter plane	100
Fly	250-300

Contraction Times of Invertebrate Muscle

Source	Time (seconds)
Anthozoan muscles	5-180
Scyphozoa	0.5-1
Earthworm Circular Muscle	0.3-0.5
Bivalve Byssus Retractor Muscle	1
Gastropod Tentacle Retractor	2.5
Horseshoe Crab Abdominal Muscle	0.195
Insect Flight Muscle (striated)	0.025 (40 beats/second)

Neural control of wing movement

- Synchronous: single volley of nerve impulses stimulates a single muscle contraction and therefore one wing stroke
- Asynchronous: only occasional nerve impulses are necessary to maintain wing movement.
- Storage of potential energy in the resilient part of the thoracic cuticle causes multiple wingstrokes after a single muscle contraction; interaction of antagonistic indirect flight muscles is also important
- Wing hinge contains Resilin, an elastic protein that returns nearly 95% of the kinetic energy that is delivered to it.
- Insects with asynchronous control generate much faster wing speeds: flies 300 beats/sec midges 1000 beats/sec; Butterflies, which have synchronous control: 4 beats/sec

- In many insects, especially those with Asynchronous control, a "click" mechanism causes wings to be stable only in the complete up or down position. Flying is more than upward-downward wing strokes.
- For effective flight the angle of the wing on the upward and the downward stroke cannot be the same.
- Direct muscles alter the angle of the wing to maintain a net thrust.
- According to the most recent studies, the insect wing stroke provides lift in three fundamentally different ways:
  - as a "classical" airfoil
  - by generating vortices behind the leading edge
  - rotational lift

Functions of wings

- Dispersal and locating of food.
- Protection against physical damage.
- Reservoir for air.
- Shield the body from excess solar radiation.
- Capture solar heat to warm the body.

8

Abdomen

- It is the posterior most and largest region of the insect-body. It is The abdomen is the metabolic and reproductive centre, where digestion, excretion, and the sexual functions take place.
- All the definitive number of segments differentiate in embryo (**Epimorphic condition**) except protura (**Anamorphosis**)
- In general more segments are visible in more generalised hemimetabolus insects than in more specialised holometabolus insects. For example in Acridids all 11 segments are visible whereas in muscidae only 2-5 segments are visible
- The basic number of abdominal segments in insect is eleven plus a **telson/ tail** (Protura) which bears anus. Abdominal segments are called uromeres.
- The undivided tergel plates overlap the sternal ones and the membranous pleura are enclosed in between them.
- The posterior part of each segment overlaps the anterior part of the segment behind it and the two segments are joined by the inter-segmental membrane.
- There are eight pairs of abdominal spiracles. The first pair lies just in front of each tympanum and the other lie dorso-laterally on each of the tergites of the segment to eighth abdominal segments.
- The terga and sterna bear antecostal ridges. The internal dorsal and ventral longitudinal muscles run one antecostal ridge to the other and act as retractors. The contraction of these muscles helps the segments to be retracted. The external muscles originate from the posterior part of the segment and attach to the anterior part of the acrotergite of the next segment. Thus, protraction of these muscles help in the extension of abdomen. Hence, musculature running longitudinally between helps in the contraction and extension of the abdomen.
- The abdominal segments are conveniently divided into three subregions i.e. pregenital, genital and postgenital segments.
    - 01-07 segments pre genital segments

- 08-09 segments genital segments
- 10-11 segments post genital segments

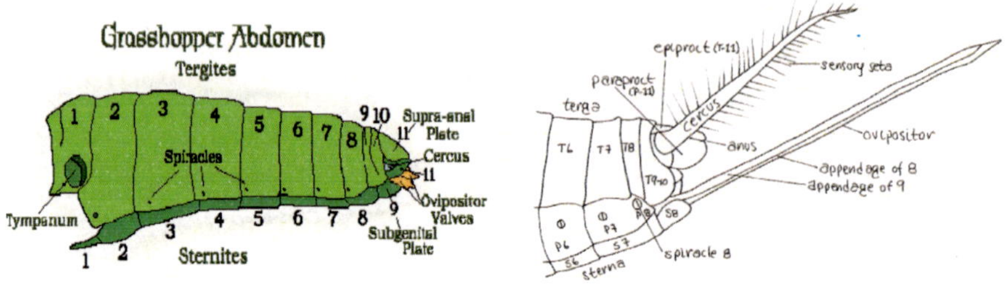

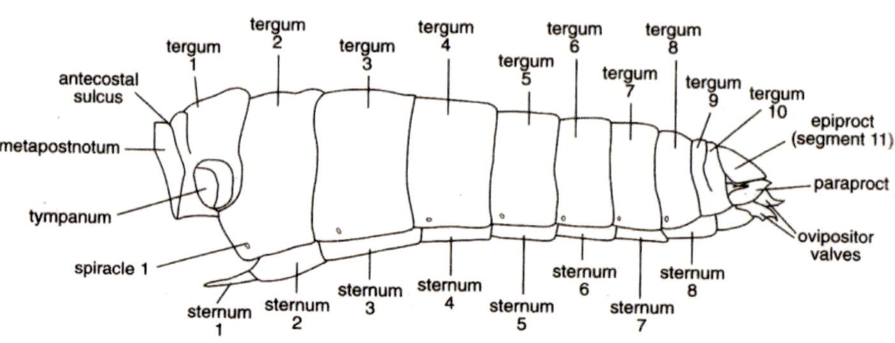

Epimorphic development	Anamorphic development
The young one after hatching from the egg consists of definite number of segments and the same number is maintained throughout. Such type of development is called Epimorphic development.	The first instar larva hatch with only 8 abdominal segments and telson. the remaining 3 segments are added at subsequent moults arising behind the last abdominal segment, but in front of the telson.
Majority of insects.	Protura

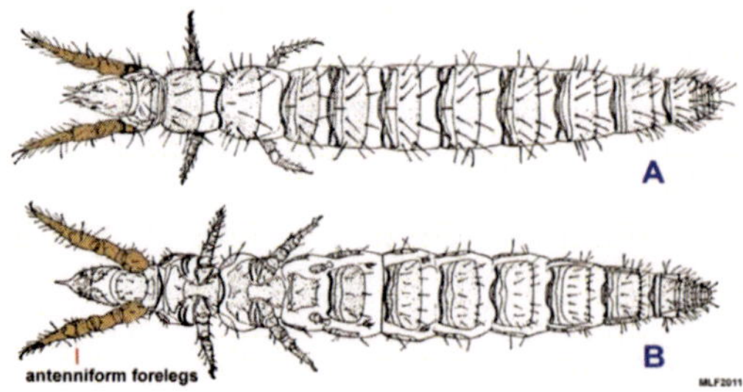

Protura

Pregenital Segments

- These segments are anterior to the position of the genital organs i.e. the segments first to seventh in female and first to eighth in male.
- The first segment is the largest and is broadly joined to the metathorax. Its tergum forms the metapostnotum and the sternum is pushed anteriority to fill the spur between meta sternellar lobes. It bears large, membrums oval tympanal organis laterally.
- The spiracles of the first abdominal segment are placed just in front of the tympanum on either side.
- In **honeybees** and wasps (Hymenoptera) the meta thoracic segment is fused with the first abdominal segment to form a structure known as **Propodeum** while the rest of abdomen structure is known as **Gaster.** The second segment of abdomen form as narrow **petiole/pedicel (waist).**
- Springtail (Collembola) has only six segments throughout their life.
- **Collembola** has special pre-genital organs :
    - Collophore or collophore or glue bar
- Present ventrally in the first abdominal segment and hence the name collembolan is derived in view of this organ. The collophore is a tube like with eversible sacs at its tip which are everted by means of blood pressure. The sticky fluid secreted from cephalic glands passes through collophore act as adhesive organ enable the insect to walk over smooth and steep surfaces. It aids in water absorption from the substratum and also in respiration.
    - Retinaculum or tentaculum or catch or hamula
    - Furcula or Furca
    - It represent the partially fused appendages found on the ventral side of the fourth abdominal segment. It enable the insect to leap in to the air suddenly and hence the common name springtails
- In protura, a pair of short cylindrical appendages on each of the first three abdominal segment and terminating in eversible vesicles.
- In order **Isoptera** (Termitidae) the queen termite after mating attains enormous proportions, this obesity is known as **Physogastry**, the increase affecting only the abdomen and not the head and thorax due to post metamorphic growth. The abdomen becomes swollen due to the enlargement of ovaries and fat bodies.

Physogastry (queen termite)

Propodeum and gaster

Abdominal Appendages in Larvae

Appendages	Remarks	Example
Abdominal gills	Filamentous or plate like 7 paired and terminated by three caudal filaments are present in the first seven abdominal segments	Ephemeroptera nymphs
Tracheal gills	Leaf like gills (lamellate) 3 numbered on abdomen	Odonata Naiads (damsel flies)
Anal papillae	A group of four papillae surrounds the anus and are concerned with salt regulation.	mosquito larvae
Dolichasters	Each dolichaster is a segmental protuberance fringed with setae.	antlion grub
Prolegs	• Two to five pairs are normally present. They are unsegmented, thick and fleshy. • The tip of the proleg is called planta upon which are borne heavily sclerotised hooks called crochets. • They aid in crawling and clinging to surface • Caterpillars: Larvae having prolegs on 3,4,5,6 and 10th abdominal segment • Semilooper : Larvae having prolegs on 5, 6 and 10th abdominal segment • Looper : Larvae having prolegs on 6 and 10th abdominal segment • Sawfly (Tenthridinidae, Hymenoptera) : Larva have 8 pair of prolegs.	larvae of moth, butterfly and sawfly

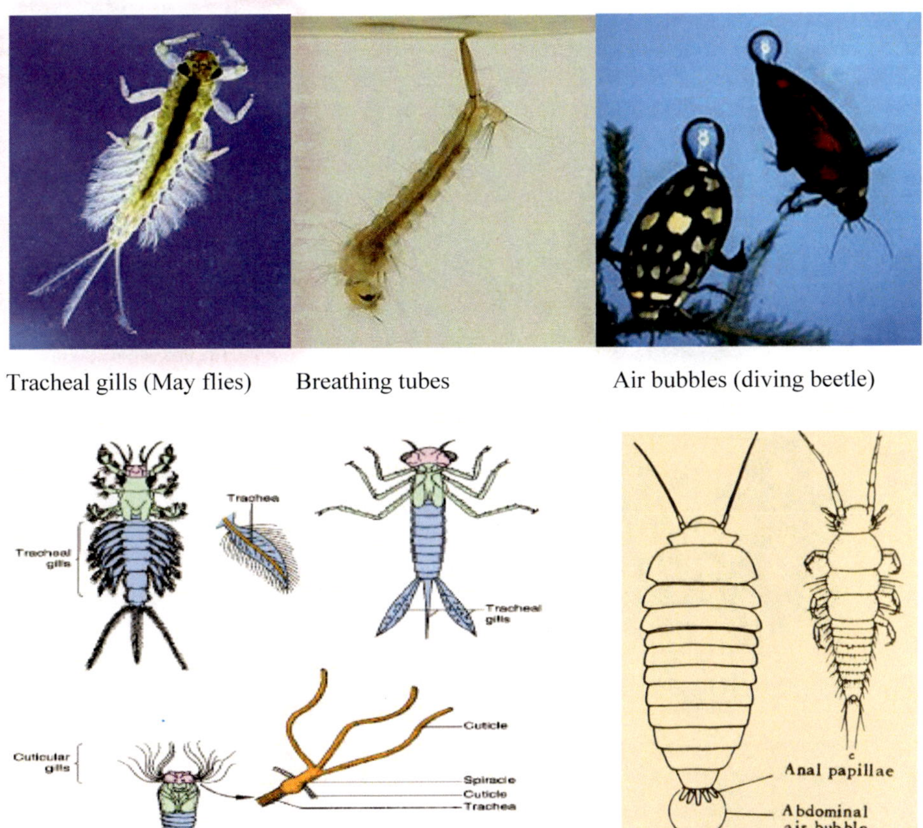

Tracheal gills (May flies) Breathing tubes Air bubbles (diving beetle)

Abdominal Appendages in Winged Adults

Sound production

- First two abdominal segment modified for sound production in Cicadidae

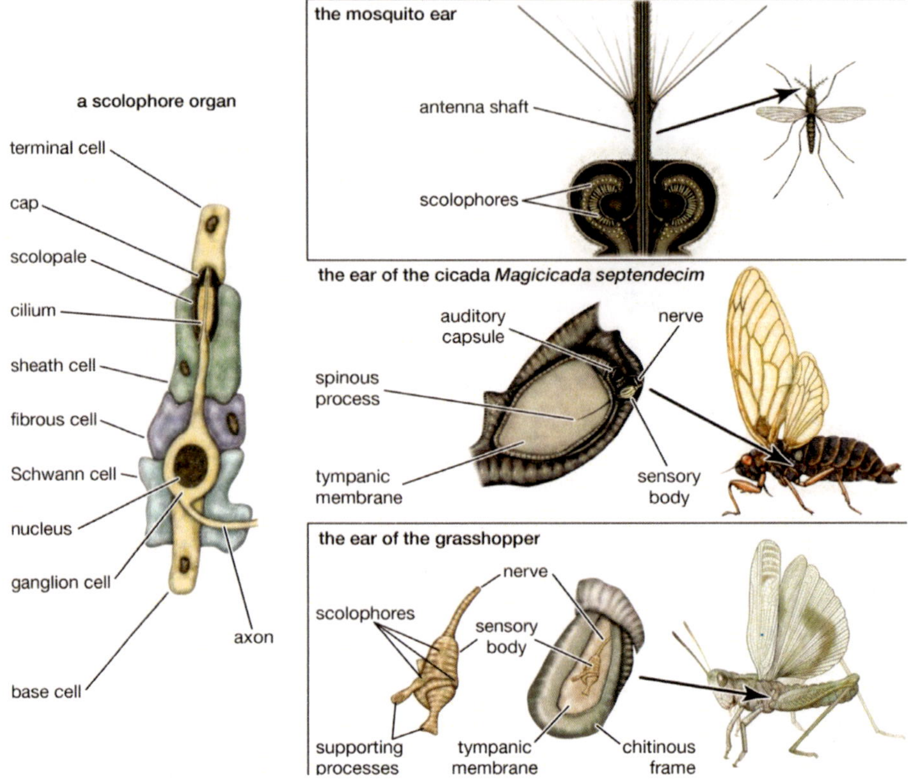

Caudal breathing tube

- It consists of two grooved filaments closely applied to each other forming a hollow tube at the apex of abdomen. e.g. water scorpion. Water scorpions (Hemiptera: Nepidae) and rat-tailed maggots (larvae of a syrphid fly). Many aquatic plants maintain their bouyancy by storing oxygen (a waste product of photosynthesis) in special vacuoles. A few insects (*e.g.* larvae of *Mansonia* spp. mosquitoes) insert their breathing tubes into these air stores and obtain a rich supply of oxygen without ever swimming to the surface of the water.

Cerci : (Cercus : Singular)

- Paired structure arised from epi and paraproct of 11th abdominal segment
- Present and well developed in apterygotes and hemimetabolus insects other than hemipteriods. In Holometabolus insects cerci varies, may be reduced or completely absent.

- Cerci function as sense organ being set with large number of trichoid sensilla with a complex articulation at base.
- These are sensitive to tactile stimuli and to air movement and sometimes act as sound receptor
- Play role in copulation
- Modify as lateral gills in larval Zygoptera
- Variably segmented cerci (1-8 segmented) : Isoptera
- Long and many segmented - e.g. Mayfly
- Short and many segmented - e.g. Cockroach
- Short and unsegmented - e.g. Grasshopper (Orthoptera), Leaf and stick insect (Phasmida)
- Sclerotised and forceps like - e.g. Earwig. Cerci are useful in defense, prey capture, unfolding wings and courtship.
- Asymmetrical cerci - Male embiid. Left cercus is longer than right and functions as clasping organ during copulation. Symmetrical 2 segmented in female embids.
- Ephemeroptera : Paired multisegmented cerci
- Absent : Hemiptera, Tubulifera (Thyanoptera), Lepidoptera

Sting

- A modified ovipositor, found only in the females of **aculeate** Hymenoptera (ants, bees, and predatory wasps).

Cornicles

- Paired secretory structures located dorsally on the dorsum of fifth or sixth abdominal segment of abdomen of aphids.
- The cornicles produce substances that repel predators or elicit care-giving behavior by symbiotic ants.

Pincers

- In Dermaptera (earwigs), the cerci are heavily sclerotized and forceps-like. They are used mostly for defense, but also during courtship, and sometimes to help in folding the wings.

Abdominal prolegs

- Fleshy, locomotory appendages found only in the larvae of certain orders (notably Lepidoptera, but also Mecoptera and some Hymenoptera).
- Prolegs are present on 3, 4, 5, 6 and 10th abdominal segment.

Abdominal gills

- Respiratory organs found in the nymphs (naiads) of certain aquatic insects.
- In Ephemeroptera (mayflies), paired tracheal gills are located along the sides of each abdominal segment; in Odonata (damselflies), the gills are attached to the end of the abdomen.

Median caudal filament

- In mayfly and silverfish, (Wingless insect) the epiproct is elongated into a circus like median caudal filament.

Pygostyles

- A pair of unsegmented cerci like structures are found in the last abdominal segment of scoliid wasp.

Anal styli

- A pair of short unsegmented structures is found at the end of the abdomen of male cockroach. They are useful to hold the female during copulation.

Ovipositor

- The egg laying organ found in female insect is called ovipositor. It is suited to lay eggs in precise microhabitats. It exhibits wide diversity and form.
- Short and horny : e.g. Short horned grasshopper
- Long and sword like : e.g. Katydid, long horned grasshopper
- Needle like : e.g. Cricket
- Ovipositor modified into sting : e.g. Worker honey bee.
- Long and permanently extruded : eg. Ichenumon fly
- Absent : Dermaptera (Earwigs), Anoplura (Sucking lice), Aphids and Coccids (Hemiptera)
- Reduced : Isoptera, mallophaga

- An appendicular ovipositor is lacking in fruit flies and house flies. In fruit flies, the elongated abdomen terminates into a sharp point with which the fly pierces the rind of the fruit before depositing the eggs.
- In the house fly the terminal abdominal segments are telescopic and these telescopic segments aid in oviposition. The ovipositor of house fly is called **pseudoovipositor** or **ovitubus** or **oviscapt**.

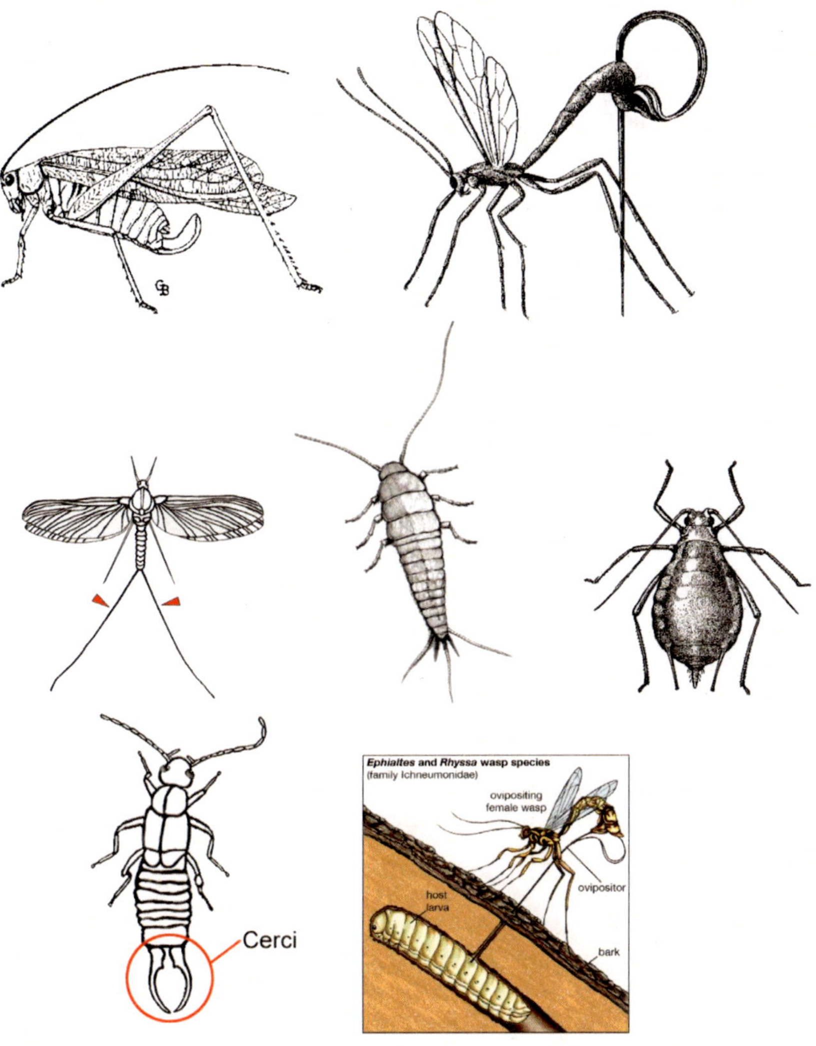

Genital Segments

- These segments bear the genitalia and differ, in the two sexes, owing to the varied peculiarities of the structures associated with them.

Female

- The eighth and ninth segments constitute the genital portion.
- The eighth segment bears the last visible sternum. The sternal plate is broad, squarish and termed the subgenital plate. The subgenital plate is produced posteriorly into one median and two lateral lobes. The median lobe is termed the egg-guide.
- The ninth segment is narrow and partially fused with the adjoining segments.
- These are the egg-laying organs of the female and are collectively termed the ovipositor. It consists of blade-like structures, termed the valves or valvula. The ventral pair of valvulae belongs to the eighth segment, the inner (= middle) and the dorsal ones to the ninth segment. The inner pair of valvulae are reduced and lie in the grooves of the ventral pair. The dorsal and ventral pairs are chisel-like and pointed upwards. The egg-guide projects between the bases of the ventral pair of valvulae.
- The gonopore of the female insect is usually situated on or behind the eigth or ninth abdominal segment.
- In **May flies** and **earwigs,** the gonopore is behind segment seven. In many orders there are no special structures associated with oviposition, although terminal abdominal segments are long and telescopic forming a type of ovipositor. Such structures are formed in lepidoptera, coleoptera and diptera.
- In **house fly**, the telescopic section is formed from segments six to nine, when not in use it is telescoped within segment five.
- In **Tephritids,** the tip of the abdomen is hardened and form a sharp point, which helps the fruit flies to insert its eggs into the fruit tissues.

Insect Abdomen - Generalized Female Genitalia

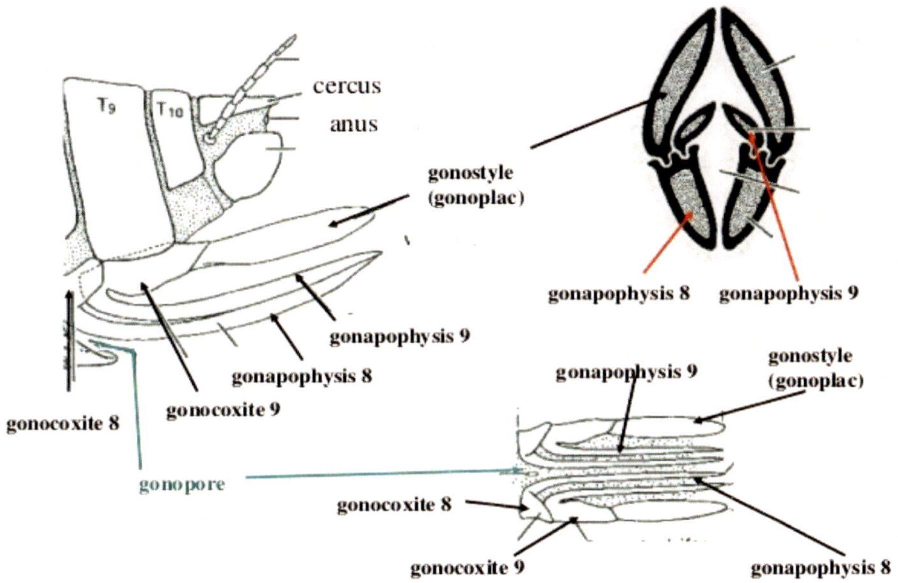

Basic forms of ovipositor

Thysanura

At the base of the ovipositor on each side are the coxae of segment eight and nine, known as first and second gonocoxae (first and second valvifers) and articulating with each of them is a slender process which curves posteriorly, known as first and second gonapophyses (valvulae) and together they form a shaft of the ovipositor. The second gogapophyses of the two sides are united, so that the shaft is made up of three elements fitting together to form a tube down which the egg passes. At the base of the ovipositor there is a small sclerite, gogangulum attached to the base of the first gonapophysis and articulate with second gonocoxa and the tergum of segment nine.

Orthoptera

An additional process is present in the second gonocoxa, known as gonoplac (third valvulae). It may or may not be a separate sclerite and may form a sheath round the gonapophyses. Gonoplac are well developed in orthoptera, where they form the dorsal valves of the ovipositor with the second gonapophyses enclosed within the shaft in tettigoniids or reduced as in gryllids. In orthoptera the gonangulum is fused with the first gonocoxa.

Hymenoptera

The first gonocoxae is absent and second gonapophyses are united. The second gonapophyses slide on the first by a tongue and groove mechanism. In symphyta and parasitic groups the ovipositor retains its original function, but in aculeate it forms the sting. Here the eggs instead of passing down the shaft of ovipositor are ejected from the opening of the genital chamber at its base. In honey bees, the first gonapophyses are known as lancet and the fused second gonapophyses as the stylet. This forms an inverted trough, which is enlarged into a basal bulb into which the reservoir of the poison gland discharges.

Male

- The ninth segment forms the genital portion.
- The basic genital elements are derived from a pair of primary phallic lobes which are present in the ventral surface of segment nine of the embryo. These phallic lobes divide to form an inner pair of mesomeres and outer parameres, collectively known as the phallomeres. The mesomeres unite to form the aedeagus, the intromittent organ. The inner wall of the aedeagus, which is continuous with the ejaculatory duct is called the endophallus, and the opening of the duct at the tip of the aedeagus is the phallotreme. The gonopore is at the outer end of the ejaculatory duct where it joins the endophallus and hence is internal. In many insects the endophallic duct is eversible and so the gonophore assumes a terminal position during copulation. The parameres develop into claspers, which are very variable in form. In many insects these basic structures are accompained by secondary structures on segments eight, nine and ten.

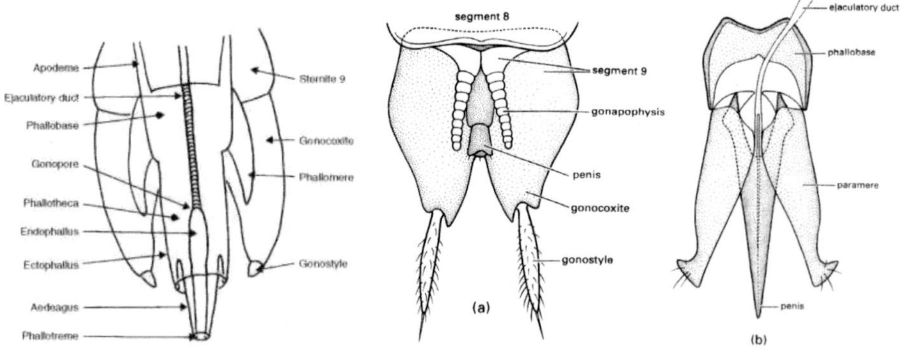

Modifications

- In collembola and dipluran, there is no intromittent organ aedeagus.
- In Thysanura, the terminal segments are similar to those in females, and sperm is not transferred directly to the female.
- In Ephemeroptera and Dermaptera- **paired penes/penis** are present.
- Claspers is derived from parameres and cerci in dermeptera and orthoptera or from paraproct in damselflies.
- In plecopteran, there is no clasper.
- In culicidae, the aedeagus lies above the anus instead of below, as a result of 180 rotation of segment eight and behind soon after eclosion.
- In Odonata, intromittent organs are present in segment two and three. Appendages which are used to clasp the female are present on segment 10, but the genital apparatus on segment nine is rudimentary.
- In Damselfly, the functional copulatory organ is present on the venter of second and third abdominal segment.
- In Ephemeroptera (Mayflies) and Dermaptera (Earwigs), the male opening is behind 7th segment.

Post Genital Segments

- The tenth and eleventh segments, in both sexes, comprise the postgenital segments.
- The tergum of the tenth segment fuses with that of the ninth on lateral sides.
- The eleventh is represented by triangular end plate, the epiproct, on above, and by paraprocts on latero-ventral sides. In between them lies a pair of conical sensory structures termed the cerci.

9

Metamorphosis

Insects development

Once an insect hatches from the egg it is usually able to survive on its own, but it is small, wingless, and sexually immature. Its primary role in life is to eat and grow. If it survives, it will periodically outgrow and replace its exoskeleton (a process known as molting). In many species, there are other physical changes that also occur as the insect gets older (growth of wings and development of external genitalia, for example). Collectively, all changes that involve growth, molting, and maturation are known as morphogenesis.

Metamorphosis

- Metamorphosis actually means change in form Metamorphosis is derived from Greek word **'Meta'** = Change, **'morph'** = form or structure.
- The changes in the development from egg to adult, collectively called Metamorphosis (life cycle).
- Post-embryonic changes in form are collectively called as metamorphosis.
- Metamorphosis refers to the way that insects develop, grow, and change form. Metamorphosis include three developmental processes namely **growth, differentiation** and **reproduction** which takes place in larval, pupal and adult stages, respectively.
- Majority of insects pass through profound metamorphosis. The growth of an immature insect is accompanied by a series of molts or ecdysis, in which the cuticle is shed off and renewed. The number of molts varies in most insects from 4-8. However some undergo as many as 20 or more molts. A few insects like Bristle tail silver fish (apterygota) may continue molting even after reaching the adult stage. But most of the insects neither molt nor increase in size once then adult stage is reached. The intervals between the ecdysis are called Stadia. The term instar is applied to form of an insect during stadium. In numbering the instar, the form assumed by the insect between hatching from the egg and the first post embryonic molt is termed "First Instar".

- Depending on the presence or absence of metamorphosis and upon the degree of metamorphosis the insects are grouped into 3 types. Ametamorphosis, Hemimeatmorphosis and Holomeatmorphosis.

Incomplete Metamorphosis

- About 12% of all insects go through incomplete metamorphosis.
- It has 3 stages.
- Eggs are often covered by an egg case which protects eggs, and holds them together. Eggs hatch into nymphs, which looks like small adults, but usually donot have wings.
- Insect nymphs eat the same food as that of adult. Nymphs shed or molts their exoskeleton (outer casings made up of chitin) and replace them with larger ones several times as they grow.
- Most nymphs molt 4-8 times. Adult stop molting when reach their adult stage. By this time, they have well developed wings instead of grown wings.

Complete Metamorphosis

- About 88% of all insects go through complete metamorphosis. It has 4 stages. Eggs are sometimes covered by hairs/scales.
- Larva hatch from eggs and do not look like adult, in any case. They usually have worm-like shape (caterpillars, maggots, grubs). Larva molts several times, and grows larger.
- Larva makes cocoons around themselves before they undergo pupation, and larva do not eat while inside cocoons. Their bodies develop in to adult with wings, legs, internal organs etc. This change takes place anywhere from 4 days to many months. Inside the cocoons, the pupa change in to adult. After a period of time, the adult breaks out of cocoon.

Ametabolous insects

- Undergo little or no structural change as they grow older. Immatures are called **young**; they are physically similar to adults in every way except size and sexual maturity. Other than size, there is no external manifestation of their age or reproductive state.

Hemimetabolous insects

- Exhibit gradual changes in body form during morphogenesis. Immatures are called **nymphs** or, if aquatic, **naiads**. Maturation of wings, external genitalia, and development of other adult structures occurs in small steps from molt to molt.

- Wings may be completely absent during the first instar, appear in the second or third instar as short wing buds, and grow with each molt until they are fully developed and functional in the adult stage.
- Developmental changes that occur during gradual metamorphosis are usually visible externally as the insect grows, but adults retain the same organs and appendages as nymphs (eyes, legs, mouthparts, etc.).

Holometabolous insects

- Have immature forms (**larvae**) that are very different from adults.
- Larvae are "feeding machines", adapted mostly for consuming food and growing in size. They become larger at each molt but do not acquire any adult-like characteristics.
- When fully grown, larvae molt to an immobile pupal stage and undergo a complete transformation.
- Larval organs and appendages are broken down (digested internally) and replaced with new adult structures that grow from **imaginal discs**, clusters of undifferentiated (embryonic) tissue that form during embryogenesis but remain dormant throughout the larval instars.
- The adult stage, which usually bears wings, is mainly adapted for dispersal and reproduction.

Hypermetamorphosis

- This is a peculiar type of development which consists of two or more types or forms of larvae in the life cycle of insects. In majority of the cases the first larval instar is **campodeiform** and the subsequent larval forms depends on type and mode of life of the larva.
- E.g.: In blister beetle (Meloidae; Coleoptera), the first larval instar is **campodeiform** followed by **scarabeiform** larval type.

Ametabola	**Hemimetabola**	**Holometabola**
No metamorphosis	Incomplete metamorphosis	Complete metamorphosis
Wingless	Wings develop during growth of young one	Wings develop during growth of adult inside pupae
Nymph	Nymph	Larva
Under go slight or no metamorphosis	Under go incomplete metamorphosis	Under go complete metamorphosis
Life cycle – 3 stages : Egg-Nymph-Adult	Life cycle – 3 stages : Egg-Nymph-Adult	Life cycle – 4 stages : Egg-Larva-Pupa-Adult

Ametabola	Hemimetabola	Holometabola
Eggs laid with no covering	Eggs often covered by an egg case	Eggs sometimes covered by hairs /scales
No changes take place during development	Young one (immature stage) is called nymph.	Young one (immature stage) is called as larvae
The development is direct (young one to adult)	The development is direct (young one to adult)	The development is indirect (young one to pupa, then to adult)
Young (immature) looks like the adult in all characters, only it may be missing sexual organs	Young one (nymph) resembles the adult in all characters except in wings	Young one (larva) does not resemble and differs from adult both in morphological characters & feeding habits.
Nymph directly becomes adult	Nymph directly becomes adult	Young one undergo pupal stage before adult stage
No pupal stages	No pupal stages	Pupal stage present
All Apterygote insects	Characteristics of lower insect orders : Exopterygote insects	Characteristic of higher orders : Endopterygote insects

The metamorphosis is regulated by endocrine system, releasing various types of hormones viz, Brain Hormone or PTTH from Corpora Cardiaca, Juvenile Hormone from Corpora Allata and Ecdysone from Pro-Thoracic Glands. The production, storage, release and function of all these growth regulating / promoting hormones are basically regulated by neurosecretary cells in brain of an insect.

Diapause in Insects

- Many species of insects have evolved a strategy called diapause.
- Diapause is a suspension of development that can occur at the embryonic, larval, pupal, or adult stage, depending on the species.
- Diapause is a physiological state of dormancy (sleep-time) with very specific triggering & releasing conditions, a neurohormonally mediated, dynamic state of low metabolic activity. Diapause is the delay in development in response to regularly and recurring periods of adverse environmental conditions.
- Animals do not grow during this time.
- Diapause is a mechanism used as a means to survive predictable, unfavourable environmental conditions, such as temperature extremes, drought or reduced food availability.
- Diapause is of two types:
- **Obligatory diapause**: It refers to the stage of suspended activity of the insect which is a hereditary character controlled by genes and is species specific. e.g. egg diapause in silkworm

- **Facultative diapause**: It is the stage of suspended activity of the insect due to unfavourable conditions and with the onset of favourable condition, the insect regains its original activity. e.g. Cotton pink bollworm *Pectinophora gossypiella.*
- The unfavourable conditions may be biotic or abiotic. Biotic conditions are natural enemies, population density etc. Whereas abiotic conditions are temperature, rainfall, humidity, photoperiod, type of food material etc.
- In some species, diapause is facultative and occurs only when induced by environmental conditions; in other species the diapause period has become an obligatory part of the life cycle.
- The latter **(obligate diapause)** is often seen in temperate-zone insects, where diapause is induced by changes in the photoperiod (the relative lengths of day and night).
- The day length when 50% of the population has entered diapause is called the critical day length, and it is usually quite sudden.
- Insects entering diapause when the day length falls below this threshold are called long day insects.
- Those insects that develop normally when there are only a few hours of sunlight and that enter diapause when exposed to longer days are called short-day insects.
- The critical day length is a genetically determined property (Danilevskii 1965; Tauber et al. 1986). Long-day insects are the ones that go into diapause because the days get shorter. Short-day insects go into diapause when there are longer days.
- Diapause is brought about by token stimuli due to change in the environment.
- Diapause begins before the actual severe conditions arise.
- Diapause during summer is called aestivation. Be active in rainy season and sleep (dormant) during drought/summer. Diapause during winter is called hybernation. Be active in summer season, and sleep (dormant) during winter.
- The difference between hibernation and estivation is that hibernations is when animals sleep during winter and live off of fat and food energy that they store for when they go into a deep sleep and estivation is when animals go into a deep sleep when is too warm

- Diapause is especially important in temperate zone insects that overwinter.
- Diapause is not only induced in an organism by specific stimuli, but once it is initiated, only certain other stimuli are capable of releasing the organism from this state. The latter is essential in distinguishing diapause as a different phenomenon from other forms of dormancy such as hibernation.
- The insects get warning signals a few times before they actually do anything about it.
- These warning signs might be: Day length, Temperature, Food etc.
- After a few warning-signal days, the female will lay 'diapausing' eggs. These eggs will have their cycle from egg to adult stopped somewhere. Some examples of these 'sleepers' are:

Insects	**Diapause stage**
Silk worm	Early embryonic stage (Egg diapause)
Locust	Mid embryonic stage (Egg diapause)
Gypsy moth	Post embryonic stage (Egg diapause)
Bamboo borer	Larval Diapause
Rice stem borer	Larval Diapause
Pink boll worm	Larval Diapause
Euproctis sp	Larval Diapause
Mosquitoes	Larval Diapause
Cabbage butterfly	Pupal Diapause
Red Hairy caterpillar	Pupal Diapause
Colorado potato beetle	Adult Diapause
Mango nut weevil	Adult Diapause

Regulation of Diapause

Diapause in insects is regulated at several levels Environmental stimuli interact with genetic pre-programming to affect the neural signaling, endocrine pathways and eventually metabolic and enzymatic changes.

Environmental

Environmental regulators of diapause generally display a characteristic seasonal pattern. In temperate regions, photoperiod is one of the most reliable cues of seasonal change. Depending on the season in which diapause occurs, either short or long days can act as token stimuli. Insects may also respond to changing day length as well as relative day length. Temperature may also act as a regulating factor, either by inducing diapause or, more commonly, by modifying the response of the insect to photoperiod. Insects may respond to thermoperiod, the daily fluctuations of warm and cold that correspond with night and day, as well as to absolute or cumulative temperature. Food availability and quality may also help regulate diapause. In the desert locust, *Schistocerca gregaria*, a plant hormone called giberellin stimulates reproductive development. During the dry season, when their food plants are in senescence and lacking giberellin, the locusts remain immature and reproductive tracts do not develop.

Neuroendocrine

Neuroendocrine system of insects consists primarily of neurosecretory cells in the brain, the corpora cardiaca, corpora allata and the prothoracic glands. There are several key hormones involved in the regulation of diapause: juvenile hormone (JH), diapause hormone (DH), and prothoracicotropic hormone. Prothoraciotropic hormone stimulates the prothoracic glands to produce ecdysteroids that are required to promote development. Larval and pupal diapauses are often regulated by an interruption of this connection, either by preventing release of prothoracicotropic hormone from the brain or by failure of the prothoracic glands to respond to prothoracicotropic hormone. The corpora allata is responsible for the production of juvenile hormone. In the bean bug, *Riptortus pedestris*, clusters of neurons on the protocerebrum called the pars lateralis maintain reproductive diapause by inhibiting JH production by the corpora allata. Adult diapause is often associated with the absence of JH, while larval diapause is often associated with its presence.

In adults, absence of JH causes degeneration of flight muscles, atrophy or cessation of development of reproductive tissues, and halts mating behaviour. The presence of JH in larvae may prevent moulting to the next larval instar, though successive stationary moults may still occur. In the corn borer, *Diatraea gradiosella*, JH is required for the accumulation by the fat body of a storage protein that is associated with diapause. *Diapause hormone* regulates embryonic diapause in the eggs of the silkworm moth, Bombyx mori. DH is released from the subesophogeal ganglion of the mother and triggers trehalase production by the ovaries. This generates high levels of glycogen in the eggs, which is converted into the polyhydric alcohols glycerol and sorbitol. Sorbitol directly inhibits the development of the embryos. Glycerol and sorbitol are reconverted into glycogen at the termination of diapause.

Tropical Diapause

Diapause in the tropics is often in response to biotic rather than abiotic factors. For example, food in the form of vertebrate carcasses may be more abundant following dry seasons, or oviposition sites in the form of fallen trees may be more available following rainy seasons. Also, diapause may serve to synchronize mating seasons or reduce competition, rather than to avoid unfavourable climatic conditions. Diapause in the tropics poses several challenges to insects that are not faced in temperate zones. Insects must reduce their metabolism without the aid of cold temperatures and may be faced with increased water loss due to high temperatures. While cold temperatures inhibit the growth of fungi and bacteria, diapausing tropical insects still have to deal with these pathogens. Also, predators and parasites may still be abundant during the diapause period. Aggregations are common among diapausing tropical insects, especially in the orders Coleoptera, Lepidoptera and Hemiptera. Aggregations may be used as protection against predation, since aggregating species are frequently toxic and predators quickly learn to avoid them. They can also serve to reduce water loss, as seen in the fungus beetle, *Stenotarsus rotundus,* which forms aggregations of up to 70,000 individuals, which may be eight beetles deep. Relative humidity is increased within the aggregations and beetles experience less water loss, probably due to decreased surface area to volume ratios reducing evaporative water loss.

10

Type of Eggs

- In most insects, life begins as an independent **egg (ovipary)**.
- Majority of insects are oviparous. Egg stage is inconspicuous, inexpensive and inactive. Yolk contained in the egg supports the embryonic development.
- Eggs are laid under conditions where the food is available for feeding of the future young ones.
- Production of eggs by the female's body is called **oogenesis** and the egg-laying process is known as **oviposition**.
- The outer protective shell of the egg is called chorion. Near the anterior end of the egg, there is a small opening called micropyle which allows the sperm entry for fertilization. Chorion may have a variety of textures.
- Size and shape of the insect eggs vary widely.

A) Singly Laid

- **Sculptured egg**
  - Chorion with reticulate markings and ridges e.g. Castor butterfly.
- **Elongate egg**
  - Eggs are cigar shaped. e.g. Sorghum shoot fly.
- **Rounded egg**
  - Eggs are either spherical or globular. e.g. Citrus butterfly
- **Nit**
  - Egg of **head louse** is called nit. It is cemented to the base of the hair. There is an egg stigma at the posterior end, which assists in attachment. At the anterior end, there is an oval lid which is lifted at time of hatching.
- **Egg with float**
  - Egg is boat shaped with a conspicuous float on either side. The lateral sides are expanded. The expansions serve as floats. e.g. *Anopheles* mosquito.

Sculptured eggs (Castor butterfly)

Elongated eggs (Sorghum shoot fly)

Rounded eggs (Citrus butterfly)

Eggs with Float (*Anopheles*)

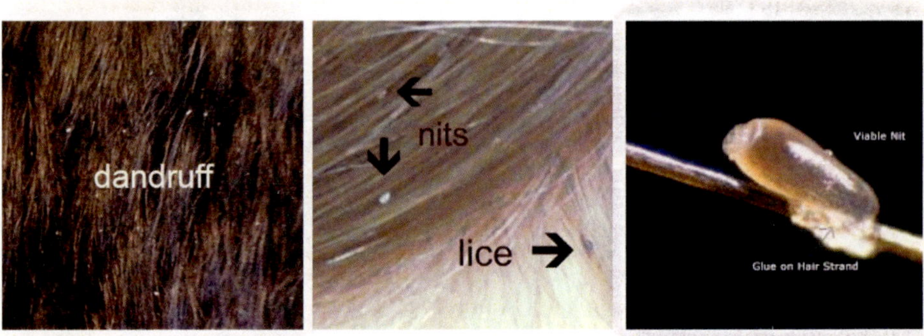

Nit (Head louse)

B) Eggs Laid In Groups

- **Pedicellate eggs**
  - Eggs are laid in silken stalks of about 1.25mm length in one groups on plants. e.g. Green lacewing fly.
- **Barrel shaped eggs**
  - Eggs are barrel shaped. They look like miniature batteries. They are deposited in compactly arranged masses. e.g. Stink bug. **Ootheca** (Pl. Oothecae)

- Eggs are deposited by cockroach in a brown bean like chitinous capsule. Each ootheca consists of a double layered wrapper protecting two parallel rows of eggs. Each ootheca has 16 eggs arranged in two rows. Oothecae are carried for several days protruding from the abdomen of female prior to oviposition in a secluded spot. Along the top, there is a crest which has small pores which permit gaseous exchange without undue water loss. Chitinous egg case is produced out of the secretions of colleterial glands.

- **Egg pod**

  - Grasshoppers secrete a frothy material that encases an egg mass which is deposited in the ground. The egg mass lacks a definite covering. On the top of the egg,the frothy substance hardens to form a plug which prevents the drying of eggs.

- **Egg case**

  - Mantids deposit their eggs on twigs in a foamy secretion called spumaline which eventually hardens to produce an egg case or ootheca. Inside the egg case, eggs are aligned in rows inside the egg chambers.

- **Egg mass**

  - Moths lay eggs in groups in a mass of its body hairs. Anal tuft of hairs found at the end of the abdomen is mainly used for this purpose. e.g. Rice stem borer.

  - Female silk worm moth under captivity lays eggs on egg card. Each egg mass is called a ***dfl*** (diseases free laying).

- **Egg raft**

  - In *Culex* mosquitoes, the eggs are laid in a compact mass consisting of 200-300 eggs called egg raft in water.

Pedicellate eggs (Green lace wing fly)

Barrel shaped eggs (Stink bug)

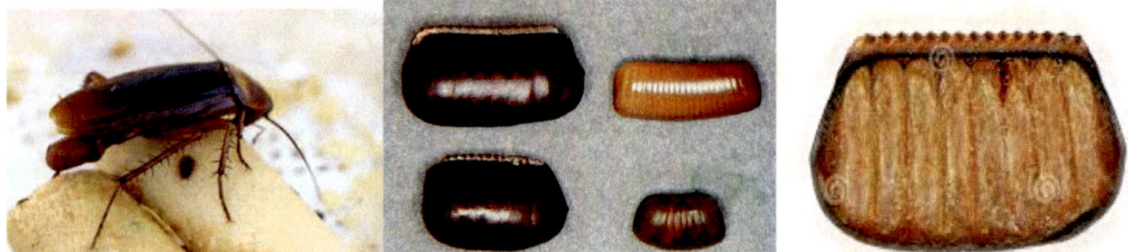

Ootheca (Cockroach)

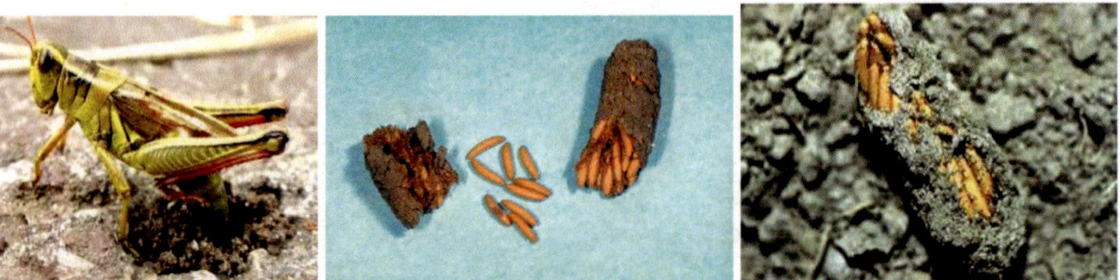

Egg pod (Grasshopper)

Egg case (Mantids)

Egg mass (Rice stem borer)

Egg raft (*Culex*)

Types of Larvae

In metamorphosis, immature stages (active feeding stages viz., nymph in hemimetabolous, larva in holo-metabolous) and pupae (non-feeding resting stage) are most important stages in metamorphosis between egg to mature stage. The nymph looks exactly like adults in most aspects except in development of wings externally, and reproductive organs internally. In all insects of hemi-metabolous type, the nymphs just like adults, and only one type is present. But, incase of holo-metabolous insects, the immature stages i.e. larva looks totally different, and larvae of various insects are typically different from others. There are several types of larvae in nature depending up on number of legs, size, shape, locomotion etc. Similarly, pupae of different group of insects look totally different from others.

This is the case, for example, in Orthoptera (crickets and grasshoppers), Hemiptera (cicadas, shield bugs, etc.), mayflies, termites, cockroaches, mantids, and Odonata (dragonflies and damselflies). Some arachnids (e.g., mites and ticks) also have nymphs. Nymphs of aquatic insects, as in the orders Odonata (dragonflies and damselflies), Ephemeroptera (mayflies), and Plecoptera (stoneflies) are also called **naiads**, which is an Ancient Greek name for mythological water nymphs, which would lure men to their deaths with their cold black hearts. In older literature, these were sometimes referred to as the heterometabolous insects, as their adult and immature stages live in different environments (terrestrial vs. aquatic). In Odonata (Ephemeroptera and Plecoptera as well) nymphs differs from adults in possessing adaptive organs like the tracheal gills, modified labium.

Leaf insect nymph

Dragon fly nymph (Naiad)

Wriggler (Mosquito)

Difference between

Nymph	**Larva**
Young one of exopterygote insects	Young one of endopterygote insects
Resemble adults and *differs from the adult with regard to the wings and genitalia*-that are present in an incompletely developed condition.	Differs from adults in appearance.
Food habits of nymph is same as adult	Food habits of larva is totally different from its adult
Wing rudiments may not be discernible in the first instar, but later become visible as wing-pads that increase in size with each instar.	No wing rudiments
Mouthparts are similar to those of adults	Mouth parts are different from adults
Compound eyes in nymphs are normal in form and function	Compound eyes are *absent* in larva and *stemmata* are present in place of compound eyes.
Growth to adult is unaccompanied by the pupal stage	Growth to adult is accompanied by the pupal stage
Nymph have 3 pairs of thoracic legs as in adults	Number of legs in larva are variable from 0-11 pairs in various orders, and legs are present in both thorax and abdomen
Wings and genital organs are underdeveloped	Wings and genital organs are totally absent
Antenna generally well developed and visible	Antenna highly reduced to 1-3 segments, always microscopic and not visible
Eg. Grasshopper, cockroach, bugs, aphids, hoppers etc.	Moths, butterfly, beetles, weevils etc.

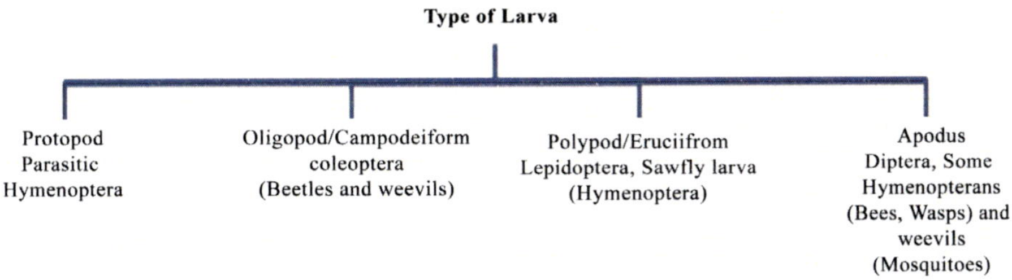

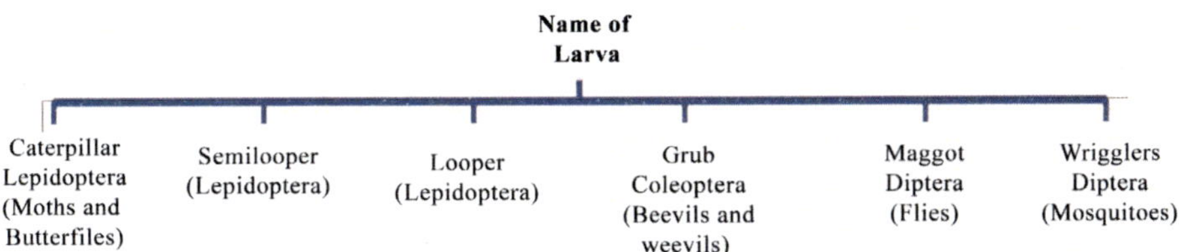

The classification of larva is mainly based on number of legs

Types of larva

- Protopod,
- Apodous,
- Oligopod
- Polypod

Protopod Larva

- Larva is pre-maturely hatched embryo (egg contains small quantity of yolk and hence larva hatches from egg before completing its embryonic development).
- Segmentation in abdomen is absent
- Rudimentary cephalic and abdominal appendages (appendages of head and abdomen are small and undeveloped).
- Nervous and respiratory systems are undeveloped.
- Cannot lead free life, and always internally parasitic on other insects.
- Examples: Parasitic Hymenoptera

Apodous Larva

- Trunk appendages or legs are completely absent.
- In place of thoracic legs, 3 pairs of sensory papillae are present.
- Apodous larvae are classified in to 3 sub types based on the development of head capsule.

Acephalous	Hemicephalous	Eucephalous
• No distinct Head Capsule. • Mouth parts represented by Mouth hook and are usually called as Maggots • Vermiform	• Appreciable reduction of the head capsule (head capsule partially developed) and its appendages, accompanied by marked retraction of the head into thorax. • Mandibles act vertically	• Head capsule more or less well developed (well sclerotized head capsule) with relatively little reduction of the cephalic appendages. • Mandibles act transversely
Cyclorrhapha (Diptera)- House flies of family Muscidae	Brachycera (Diptera)- Horseflies and Robberflies of family Asilidae	Nematocera (Diptera)- Mosquito of family Culicidae (Wriggler) and grub of red palm weevil.
	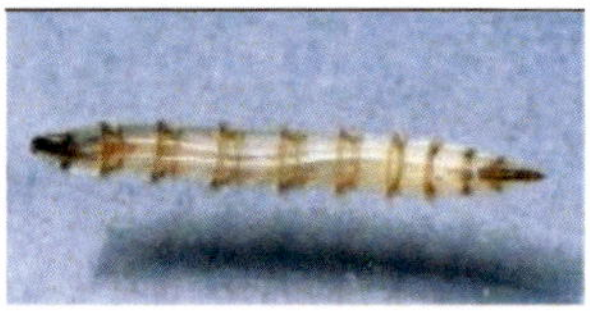	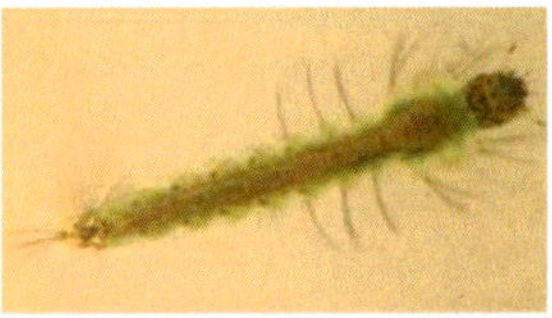

Oligopod Larva

- More or less developed thoracic legs
- Well developed head capsule and its appendages
- No abdominal appendages (abdominal legs absent)

Oligopod larvae are classified in to 2 types based on the shape and other characters.

Campodeiform

- They are so called from their resemblance to the dipluran genus Campodea.
- Body is elongate, depressed dorsoventrally and well sclerotised. Head is prognathous. Thoracic legs are long.
- A pair of abdominal cerci or caudal processes is usually present.
- Larvae are generally predators and are very active.
- e.g. grub of antlion or grub of lady brid beetle.

Scarabaeiform

- Body is `C' shaped, stout and subcylindrical.
- Head is well developed.
- Thoracic legs are short.
- Caudal processes are absent.
- Larva is sluggish, burrowing into wood or soil.
- e.g. grub of rhinoceros beetle.

Difference between

Campodeiform	Scarabaeiform
Straight body (Allegator like)	C shaped body
Body is flat dorso-ventrally compressed	Body is cylindrical or sub-cylindrical
Body well scleortized	Body wall soft and fleshy
Head prognathous	Head hypognathous
Thoracic legs relatively well developed and long (body looks like thrips)	Thoracic legs relatively reduced, short
Pair of anal cerci or stylus may be present	Anal cerci absent
Usually active	Inactive
Mostly predators	Mostly phytophagous
Neuroptera, Trichoptera, Coccinellid beetles, Ground beetles	Scarabaeidae family of Coleoptera (dung rollers, root grubs, rhinoceros beetles)

Campodaeiform larva

Polypod Larva (Eruciform larva)

- The body consists of an elongate trunk with large sclerotised head capsule.
- Head bears a pair of powerful mandibles for phytophagy
- Two groups of single lensed eyes (Stemmata) found on either side of the head constitute the visual organs.

- Well developed segmentation in thorax and abdomen
- Antenna present but very small in size or rudimentary
- Both thoracic (three pairs) and abdominal prolegs/pseudolegs (2-5 pairs) are present (appendages of head and abdomen are well developed)
- Pseudolegs are segmented and they end in claws which are used for holding on to the leaf. Bottom of the proleg is called **planta** which typically bears rows or circlet of short hooked spines or **crochets** which are useful in clinging to the exposed surface of vegetation and walking.
- Peripneustic respiration is common
- Generally Phytophagous
- The larvae of lepidoptera (butterflies and moths) are called **caterpillars**
- The larvae of Coleoptera and hymenoptera are called **grubs**
- The larvae of diptera are called **maggots**
- The larvae of mosquitoes are called **Wriggler**
- Polypod larvae can be classified in to various types based on the number of legs, locomotion and other structural characters.
- Examples: Some Lepidoptera and Sawflies of Hymenoptera

Caterpillar **Semilooper** **Looper**

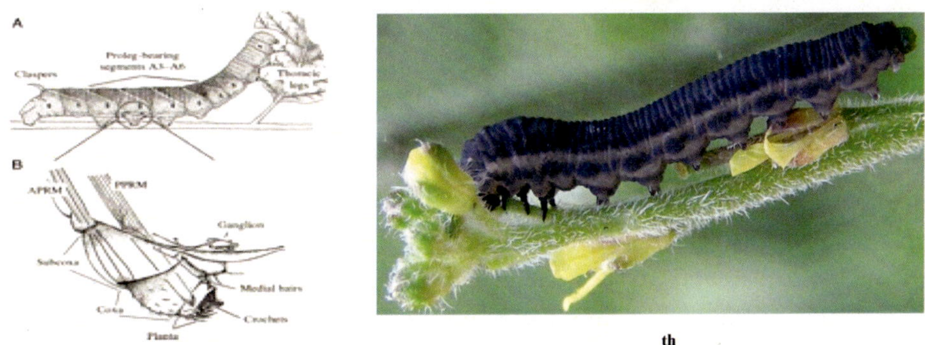

Mustard sawfly larva (8 pairs of prolegs on 1-8th abdominal segments

Caterpillar

- Larva have 5 pairs of prolegs on 3rd, 4th 5th , 6th and 10th abdominal segment.
- Cabbage butterfly, *Spodoptera litura, Helicoverpa armigera*
  - **Hairy caterpillar**- Hairs on body are dense, sparse or arranged in tufts and may cause irritation, when touched. e.g. Red hairy caterpillar, *Spilosoma obliqua,* Red Hairy Caterpillar *Amsacta albistriga,* Castor Hairy Caterpillar *Euproctis lunata, E. similis*
  - **Slug caterpillar** - Larva is thick, short, stout and fleshy. Laval head is small and retractile. Thoracic legs are minute. Abdominal legs are absent. Abdominal segmentation is indistinct. Larva has poisonous spines called **scoli** distributed all over the body. Such larva is also called **platyform** larva.
  - **Elateriform** - Body long, smooth, and cylindrical with hard exoskeleton and very short thoracic legs. E.g. Wireworm larva, click beetle, flour beetle.
  - **Sphingid Larva (Horn worm)** : Posses well developed *hook like structure on the dorsal surface of 8th abdominal segment*, in addition to all polypod characters. Hence these known as *Hornworms*. Example: Sphinx moth (death head moth)-*Acherontia spp.*

Semilooper

- First two pairs of prolegs (3rd and 4th abdominal seg) are reduced. Due to reduction of size of first prolegs, during locomotion the body form a loop shape, and hence these are called semiloopers.
- Castor Semilooper *Achaea Janata*

Looper

- Generally only two pairs of prolegs (6th and 10th abdominal segment) are present and other prolegs are absent, and hence during the locomotion, the larva gives complete loop like shape, and hence these are called *loopers*. Cabbage Looper, *Trichoplusia ni*

Type of Pupae

Insects emerge (**eclose**) from pupae by splitting the pupal case, and the whole process of pupation is controlled by the insect's hormones. Most butterflies emerge in the morning. In mosquitoes the emergence is in the evening or night. In fleas the process is triggered by vibrations that indicate the possible presence of a suitable host. Prior to emergence, the adult inside the pupal exoskeleton is termed "**pharate**". Once the pharate adult has eclosed from the pupa, the empty pupal exoskeleton is called an "**exuvium**" (or **exuvia**); in most hymenopterans (ants, bees and wasps) the exuvium is so thin and membranous that it becomes "crumpled" as it is shed.

- It is the resting and inactive stage in all holometabolous insects. During this stage, the insect is incapable of feeding and is quiescent. During the transitional stage, the larval characters are destroyed and new adult characters are created.
- Pupa is the resting, inactive stage of the holometabolous insects and transitional phase during which the wings are developed and the insect attain matured sexual organs.
- Pupae are inactive and usually sessile (not able to move about) except in some cases where they crawl (Neuroptera) (Aphid lion), can swim e.g.: mosquitoes.
- The pupation may takes place either in soil, or on the plant surface or within the webs. They have a hard protective coating and often use camouflage to evade potential predators.
- The pupal stage is found only in holometabolous insects, those that undergo a complete metamorphosis, going through four life stages; embryo, larva, pupa and imago.
- Pupae may further be enclosed in other structures such as cocoons, nests or shells.
- In the life of an insect the pupal stage follows the larval stage and precedes adulthood (imago). During pupation, the adult structures of the insect are formed whilst the larval structures are broken down.
- The duration of pupation may vary from weeks (2 weeks as in monarch

butterflies), months or even years. E.g. Anise Swallowtails sometimes emerge after years as a chrysalis.

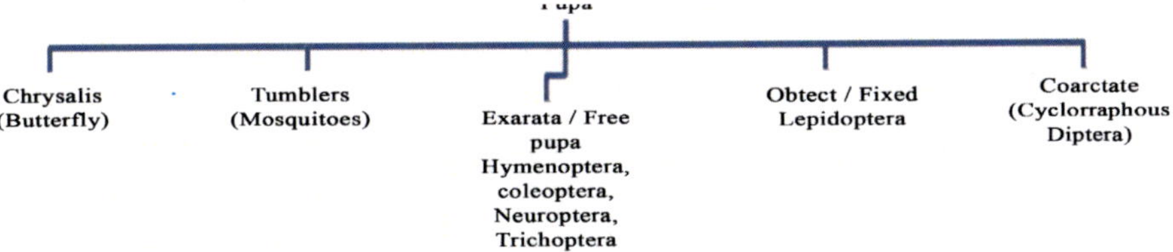

Pupae is divided on the following bases

Based on the presence or absence of **powerful mandibles**

Decticous Pupa

- Posses relatively powerful sclerotized articulated mandibles used to escape the adult from cocoon or cell.
- They are primitive type and always exarate (free appendages, not adhering to rest of body).
- Examples: Neuroptera, Mecoptera, Trichoptera.

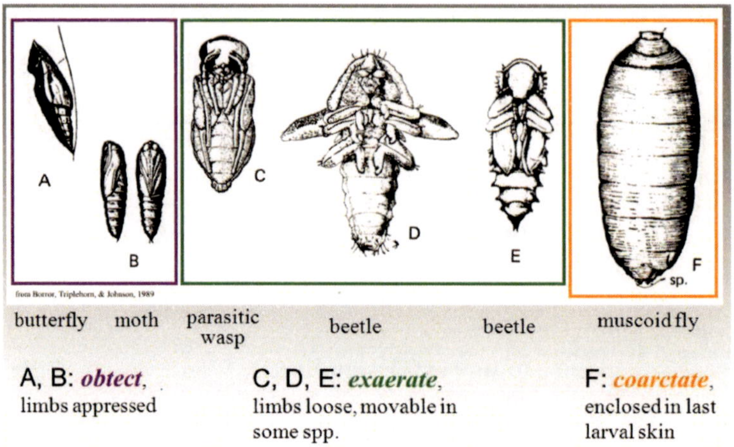

A B C D E F sp.

from Borror, Triplehorn, & Johnson, 1989

butterfly moth parasitic wasp beetle beetle muscoid fly

A, B: ***obtect***, limbs appressed

C, D, E: ***exaerate***, limbs loose, movable in some spp.

F: ***coarctate***, enclosed in last larval skin

Exarate

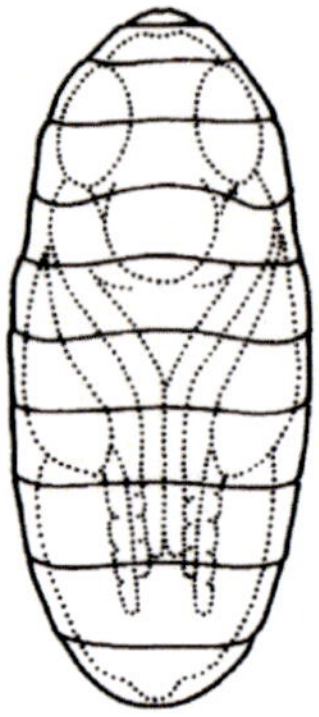

Coarctate

Obtect

Coarctate (Puparium)

Chrysalis

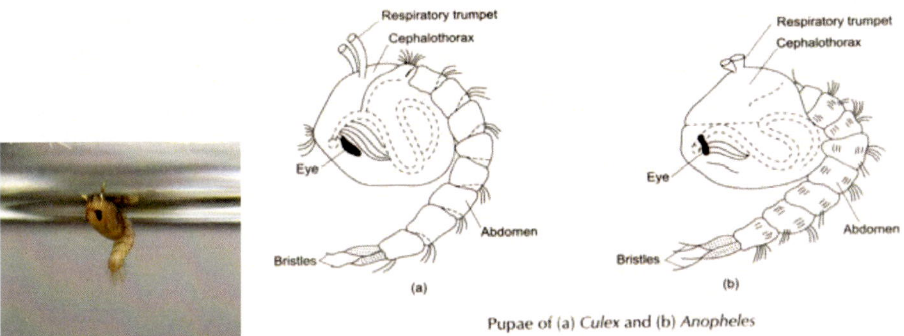

Pupae of (a) *Culex* and (b) *Anopheles*

Tumbler Pupa

Adecticous Pupa

- Having non-articulated often reduced mandibles that in most species are not used for escape from the pupal cocoon.
- Examples: Lepidoptera and Diptera.

Difference between

Decticous Pupa	Adecticous Pupa
Possess relatively powerful mandibles which are used for escaping of the adult from the cocoon i.e. to break the cocoon. e.g.: Neuroptera	Do not possess the mandibles but with the help of other appendages, adults escape from the cocoon eg: Lepidoptera, Diptera.

Based on the shape and attachment of appendages, adecticous pupae are classified int following types:

1. Obtect

- Adecticous type.
- Various appendages of the pupa viz., antennae, legs and wings pads are glued to the body by a secretion produced during the last larval moult.
- Exposed surface of the appendages are more heavily sclerotised than those adjacent to body.
- Eg: moth pupa. All the moths belong to Lepidoptera

2. Chrysalis

- Adecticous, obtect type
- The pupa is angular and attractively coloured.
- The pupa is attached to the substratum by hooks present at the terminal end of the abdomen called **cremaster**.
- The middle part of the chrysalis is attached to the substratum by two strong silken threads called **gridle**.
- E.g, All butterflies belong to lepidoptera

3. Tumbler

- Obtect type
- It is comma shaped with rudimentary appendages.
- Breathing trumpets are present in the cephalic end and anal paddles are present at the end of the abdomen.
- Abdomen is capable of jerky movements which are produced by the anal paddles.
- The pupa is very active.
- E.g. : Pupa of mosquito

4. Exarate

- Adecticous type
- Various appendages viz., antennae, legs and wing pads are not adhering to the body.
- All wings, legs, antenna, mouth parts are independent, not attached except at their point of origin.
- All oligopod larvae will turn into exarate pupae.
- The pupa is soft and pale
- e.g. Pupa of rhinoceros beetle. Examples: Most of Coleoptera, Diptera, Hymenoptera, Primitive Lepidoptera.

5. Coarctate

- Adecticous Exarate type
- The pupa remain closed in a puparium which is formed from the preceding/ last larval cast skin and pupa looks like a capsule or a barrel.

- The pupal case is barrel shaped, smooth with no apparent appendages.
- The hardened dark brown pupal case is called puparium.
- Examples: House flies (Diptera), Suborder Cyclorrapha

Pupal Protection

In general pupal stage lacks mobility. Hence it is the most vulnerable stage. To get protection against adverse conditions and natural enemies, the pupa is enclosed in a protective cover called cocoon. Based on the nature and materials used for preparation of cocoons, there are several types.

Type of cocoon	**Materila used**	**Example**
Silken cocoon	Silk	Silk worm
Earthern cocoon	Soil and saliva	Gram pod borer
Hairy cooccon	Body hairs	Wooly bear
Frassy cocconn	Frass and saliva	Coconut black headed caterpillar
Fibrous cocoon	Fibres	Red palm weevil
Puparium	Hardened last larval skin	House fly

11

Digestive System

Alimentary canal of an insect is a tubular structure passing through the body that provides the special internal environment for the food to be mechanically and chemically disintegrated and brought to the vicinity of absorbent cells. Hence, it is involved in the transport of nutrients to the individual cells. It is concerned with ingestion, trituration (chewing), digestion, absorption of nutrient material into haemolymph and egestion of unused food materials. It is tubular structure, which extends from mouth to the anus. It is divided into three distinct regions, namely fore-gut (stomodaeum), mid-gut (mesenteron) and hind-gut (proctodaeum). The fore- and hind-guts are ectodermal in origin and the mid-gut is endodermal.

Longitudinal and circular (intrinsic) muscles are usually associated with each of these three regions and by means peristaltic contractions move the food along the alimentary canal. The functional stomach is the mesenteron usually a circular valve-like fold separates the cavities between stomodaeum and mesenteron is known as stomodael or cardiac valve, and the one closing the entrance to the proctodaeum as the proctodaeum or pyloric valve.

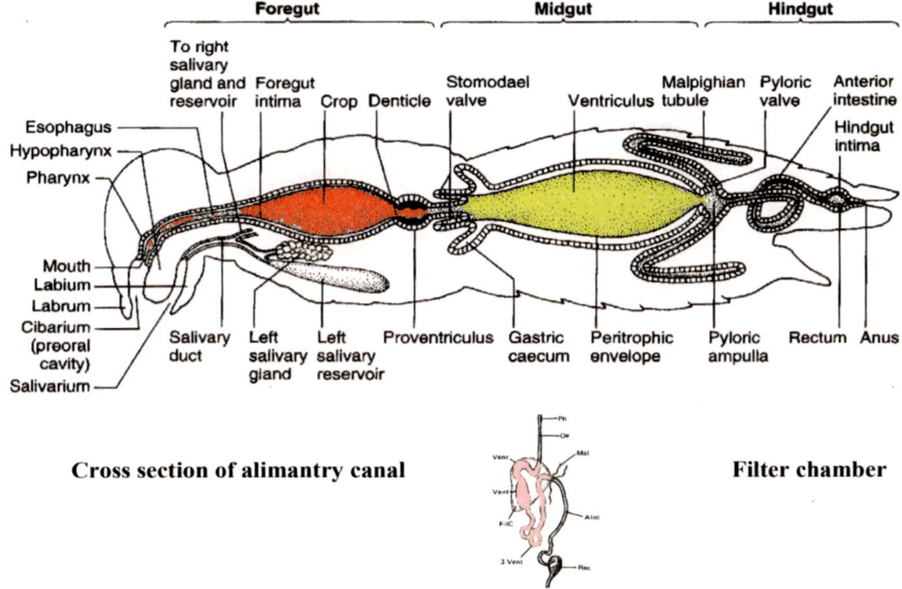

Cross section of alimantry canal

Filter chamber

Fore-gut

An insect's mouth, located centrally at the base of the mouthparts, is a muscular valve (sphincter) that marks the "front" of the foregut. Food in the buccal cavity is sucked through the mouth opening and into the pharynx by contractile action of cibarial muscles. These muscles, located between the head capsule and the anterior wall of the pharynx, create suction by enlarging the volume of the pharynx (like opening a bellows). This "suction pump" mechanism is called the cibarial pump. It is especially well-developed in insects with piercing/sucking mouthparts.

It is ectodermal in origin and lined inside by a layer of cuticles, known as the intima, which is shed at each moult in the same way as rest of the cuticle. The intima is impermeable; hence no secretion and absorption take place in the fore-gut. Histologically the following layers, passing from within outwards, are generally recognizable in the walls of the fore-gut. They are:

- The intima or innermost lining, which is cuticular layer directly continuous with the cuticle of the body wall.
- The epithelial layer, continuous with the epidermis and secreting the intima, it is often extremely thin.
- The basement membrane bounding the outer surface of the epithelium.
- The longitudinal muscles
- The circular muscles
- The peritoneal membrane, which consists of apparently structure-less connective tissue and is often difficult to detect small amounts of connective tissue, may also lie among the muscle.

Fore-gut is concerned with the conduction and storage grinding and mixing, and a limited amount of digestion. it is divisible into the following parts:

Pharynx

- It is the first part of the fore-gut following on from the preoral cavity.
- It lies between mouth and oesophagus.
- The cibarial and pharyngeal dilator muscles help to pull in and push the food backwards through the oesophagus into the crop.
- The pharynx may be highly modified into a pump seen in sucking insects like Lepidoptera, Hemiptera, and many Diptera.

Oesophagus

- It is a tubular, undifferentiated structure between pharynx and crop.
- It passes from the hinder region of head into the fore part of the thorax.

Crop

- It is the enlarged, sac-like posterior portion of the oesophagus meant for storing food.
- Frequently the crop is folded longitudinally and transversely when empty, becoming distended when the insect feeds.
- A limited amount of digestion of stored food may take place by the action of salivary enzymes and the enzymes which are regurgitated from the mid-gut. The secretion and absorption is prevented by a layer of impermeable intima.
- Some groups have a modified crop that is distensible (Lepidoptera, Hymenoptera, and Diptera) allowing for additional storage.

Proventriculus/Gizzard

- It lies immediately posterior to the crop and has six powerful muscular ridges to break (to grind) the food materials into smaller particles.
- The stomodael valve (foregut valve) regulates the flow of material from the foregut to the midgut and generally demarks the boundary between the two. It controls the onward passage of food to the mid-gut and also prevents regurgitation of food from the mid-gut to the fore-gut.
- Fluid-feeding insects lack a proventriculus.

Mid-gut

- It is termed the functional stomach or ventriculus or mesenteron. It typically begins with the cardiac epithelium associated with the stomodael valve. Immediately posterior to the stomodael valve, there is commonly a group of diverticula, the gastric caeca. The remainder part of the mid-gut is the ventriculus.
- Histologically the walls of the stomach exhibit the following structure. Internally it is lined by the mid-gut epithelium, the outer ends of whose cells rest upon the basement membrane and internally possess the peritrophic membrane.
- The basement membrane is followed by an inner layer of circular muscles and, outer layers of longitudinal muscles and outermost of mid-gut is a thin peritoneal membrane. Both the muscular layers are composed of striated fibres and their positions are the reverse of those in the fore-gut.
- Mid-gut is endodermal in origin and lined by peritrophic membrane, which is semipermeable, extremely thin and consists of protein-like (muco-protein) cuticle, permits easy passage of liquids and solutes for absorption.

it protects the epithelial cells from abrasion by the food particles. Mid-gut is concerned with the secretion of enzymes, digestion and absorption of the products of digestion.

Gastric caeca

- These are six, large, fleshy diverticula at the anterior end of the mid-gut and each caecum bears a long anterior and a short posterior lobe.
- They help to increase the surface area of the mid-gut for absorption. They appear to be active in lipid absorption.

Hind-gut

- it is the posteriormost region of the alimentary canal, ectodermal in origin and lined by relatively thin intima. Histologically, it consists of the same layers as the fore-gut except that its circular muscles are developed both inside and outside the layer of longitudinal muscles.
- The hind-gut ejects the unabsorbed and undigested food posteriorly through the anus. The Malpighian tubules attach to the pylorus region of the hindgut. The commencement of this hind-gut is normally marked by a pyloric or proctodael valve and the insertion of malpighian tubules. The region is termed the pylorus.
- In aphids the posterior part of the hindgut is in contact with the anterior part of the foregut to remove fluid from the body rapidly. This fused portion is known as filter chamber.

Ileum

- It is the anterior most wide portion of the hind-gut.

Colon

- It is the middle, narrow and with a s-shaped bend. The bend helps to break the peritrophic membrane enclosing the food material.

Rectum

- It is bulb-like, bears six rectal pads or papillae whose intima is permeable.
- They absorb water, salts, sugars and amino acids from the outgoing residues.
- Efficient recovery of water is facilitated by six rectal pads that are embedded in the walls of the rectum.
- These organs remove more than 90% of the water from a fecal pellet before it passes out of the body through the anus.

Molting

- The foregut and the hindgut are lined with a chitinous unsclerotized layer that is shed during molts.
- The gut epithelial cells are continuous with the epidermal cells.
- Before a molt, the insect takes in air or water into the digestive system to increase its internal hemolymph pressure allowing for the rupture of the cuticle.

Salivary glands

- These are the paired, labial glands which secrete saliva. the single, medium salivary duct opens in the salivarium between the labium and hypopharynx.
- The glandular cells extend back into the thorax like clusters of grapes. These structures usually reside in the thorax (adjacent to the foregut). Salivary ducts lead from the glands to the reservoirs and then forward, through the head, to an opening (the salivarium) behind the hypopharynx. Movements of the mouthparts help mix saliva with food in the buccal cavity.
- They produce saliva which serves to lubricate the mouthparts and contain enzymes which digest the food materials. The enzyme, amylase converts starch to sugar and invertase converts sucrose to glucose and fructose.
- In Lepidopteran caterpillars and Caddisfly larvae they have been converted to the production of silk, while in the Queen Honey Bee they are called the mandibular glands and secrete hormones. In Mosquitoes, these produce anticoagulant.

Digestion and absorption of food

Green plants are autotrophs. Insects, however, are heterotrophs. In addition to carbohydrates, fats, proteins and nucleic acids, insects also acquire water, vitamins, and minerals from their food. Most terrestrial insects are highly-adapted for water conservation and get most of the water they need directly from their food.

The digestive system is responsible for breaking down foods into their component parts:

carbohydrates ---> monosaccharides

proteins ---> amino acids

fats ---> glycerol and three fatty acids.

These processes begin in the foregut and are facilitated by salivary enzymes (such as amylase).

a) Foregut -- physically and developmentally part of the exoskeleton; cuticle lined; site of food breakdown.

(i) proventriculus -- responsible for physically grinding food and closing off the foregut and the midgut.

(ii) crop -- food storage.

b) Midgut -- not cuticle lined, but covered internally by a thin peritrophic membrane which is composed of cuticle. Peritrophic membrane protects the inner surface of the gut wall from abrasion and also serves as a barrier to pathogens that may enter via the digestive tract.

c) Hindgut -- part of exoskeleton; cuticle lined; site of water resorbtion; plays important role in excretion

Though insects possess a large number of digestive enzymes, they are often helped by the presence of symbiotic micro-organisms, such as protozoa in the case of the termites and some primitive cockroaches which feed on wood, and bacteria in the wax moth *Galleria mellonella* which feeds on the wax that honey bees uses to make the combs in its hives.

Excretory System

It is concerned with the maintenance of constant internal environment with proper level of salts and water, and osmotic pressure in the haemolymph by the elimination of toxic nitrogenous wastes derived from the metabolism of proteins. It plays a key role in physiological homeostasis. In insects the main excretory organs are Malpighian tubules.

Components of the excretory system

- Malpighian tubules
- Fat bodies
- Nephrocytes
- Labial kidney
- Digestive tract

Malpighian Tubules

- These are discovered by an Italian Anatomist, Marcello Malpighi, in 17th century (1669) and have been named after him by Meckel in 1829.
- These are long, slender, yellowish tubes closed at their distal ends and open in the pylorus near the junction of midgut and hindgut. They lie freely in the body cavity and are bathed directly by the haemolymph.
- The number of Malpighian tubules is variable. They occur in twos, or

multiples of two, and their primitive number is six. Their number varies from 4 (Hemiptera and Thysanura) to 200 (Odonata and Orthoptera).

- Aphids and Collembola lack Malpighian tubules altogether.
- Vestigial MT present in Protura, Strepsiptera and Diplura.
- Main function: remove nitrogenous (metabolic) wastes; main excretory product is uric acid (>80%).

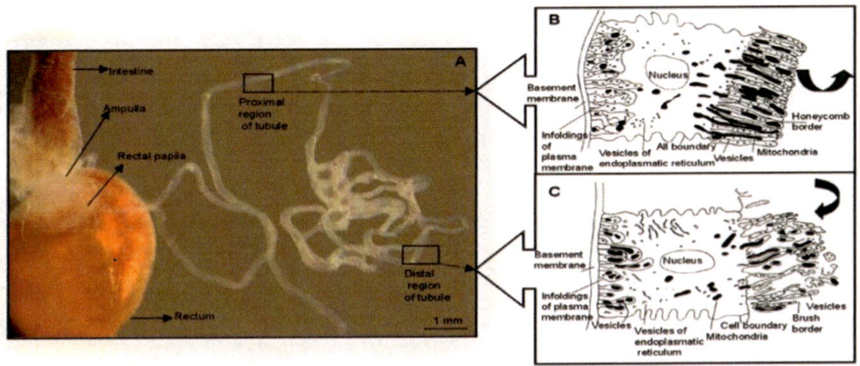

Excretory System of *Rhndniua pmlixus*. **A)** Excretory system and the posterior part of the alimentary canal. In evidence, it is shown only one complete Malpighian tubule; **B)** Schematic diagrams of a cell from the distal region of a Malpighian tubule showing the regular cytoplasmic filaments of the honeycomb border; **C)** Schematic diagrams of a cell from the proximal region of a Malpighian tubtile showing the irregular filaments ofthe brush border. Arrows indicate directional flux of secretion on Malpighian tubule (modified from Wigglesworth and Salpeter 1962).

Structure

- Histologically they resemble the wall of mesenteron rather than that of proctadaeum.
- Malpighian tubules are the main excretory organ. Blind-ended tubules ~ 100 μ dia.; single layer of epithelial cells; 2-70 mm in length.
- Total surface area of all tubules is proportional to body size: 0.54 + 0.22 mm^2/mg body wt
- The wall of MT is composed of 4-6 layers of epithelial cells having 3 regions

  a. a basal one filled with vacuoles, mitochondria and other intracellular components

  b. a middle region with nucleus

  c. an apical region with brush border, two types :

i. the honeycomb border containing a number of closely packed, parallel filaments or microvilli

ii. Typical brush border with separate independent filaments

- Internally the cells produce a closely packed microvilli called the honey comb border in the distal 2/3 part, and in the basal region (proximal 1/3) endoplasmic reticulum is developed which forms the brush border. These are the main secretory cells of the tubule.
- The distal region is concerned with absorption of water and other material from haemolymph
- The proximal part is concerned with the re-absorption of water and useful materials particularly Na and K.
- Epithelial cells rest on basement membrane (*Tunica prophria*) which is covered by a peritoneal cost often containing muscle fibres. These muscles produce peristaltic movement of the tubule.

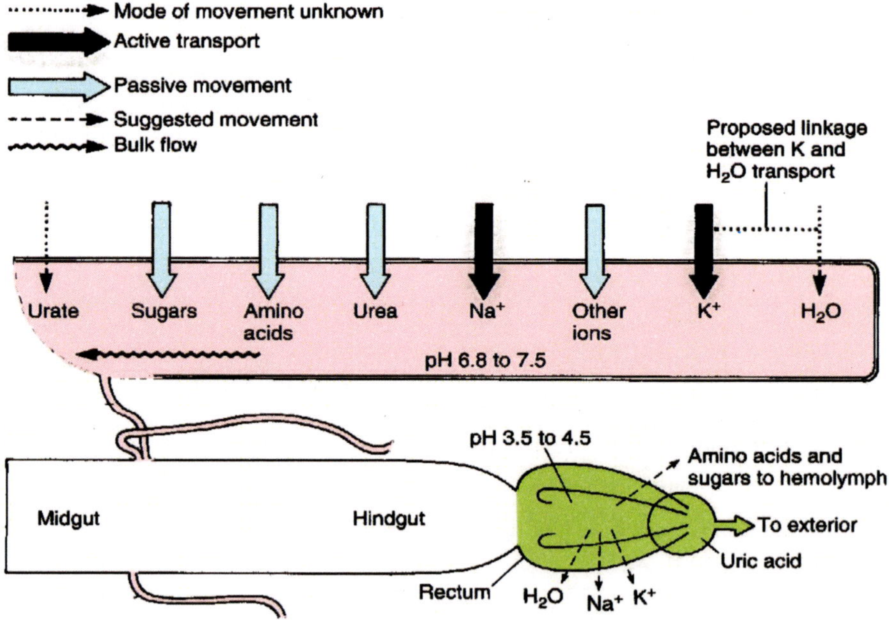

Types of Malphigian tubules

a. Fundamental / Primitive type

- MT are long and convulated tubes lying free in the body cavity (orthoptera) or may be blind or attached to and lying with in the tissues surrounding the hindgut

b. Coleoptreous type

- Mt are attached distally to one side of the gut or may be disposed of around the rectum in a simple convulated layer (Cryptonephridial arrangement).

c. Hemipteran type

- Tubes are free at distal ends but the proximal half is opaque and the distal half is clear

d. Lepidopterous type

- Cryptonephridia are also present, but like hemipteran type, the proximal half is opaque and the distal half is clear.

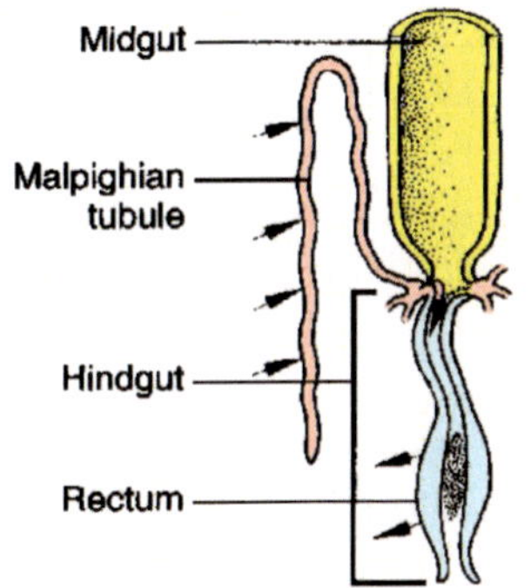

Fundamental or Primitive type

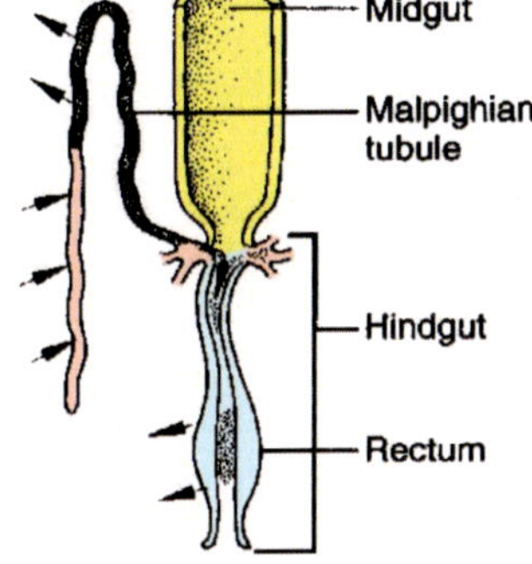

Hemipteran Type

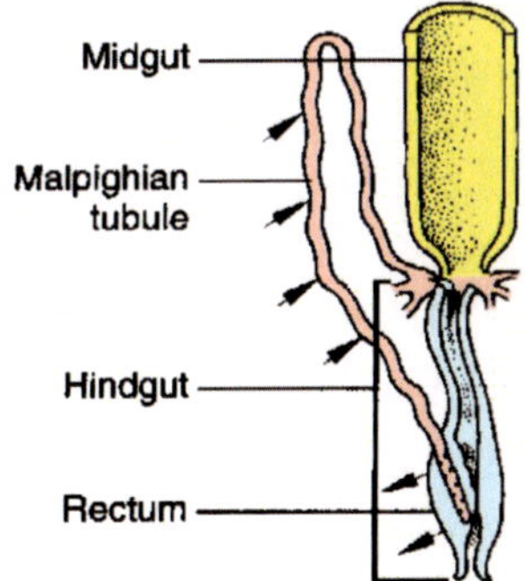

Coleoptreous type (Cryptonephridial type)

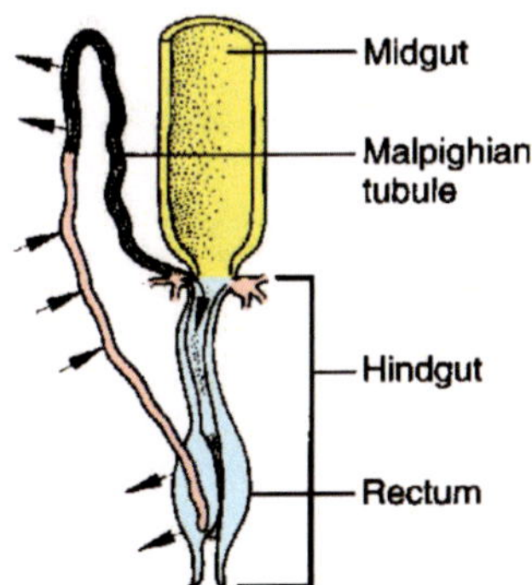

Lepidopterous type

Storage Secretion

- In insects without MTs (e.g Collembola), uric acid granules collect in the urate cells present in the fat body. As the insect become older, these urate cells become loaded with the granules. This process of excretion of the deposition of the excess products of metabolism is called Storage excretion.
- Because uric acid insoluble can be stored without adverse effects on physiology; some cockroaches store uric acid hemolymph or in specialized **urate cells** in fat body – can utilize store in times of stress. Also male *Blatella germanica* store it in accessory glands material transferred to female during copulation.
- In some Lepidoptera, fat body changes from excreting uric acid to storing it. Occurs before pupation. Stores transported to rectum shortly before eclosion (and eliminated as meconium).

Labial kidney

Collembola and thysanura possess the tubular glands opening above the base of labium by a duct called as Labial kidney.

Nephrocytes

- Scattered /localised groups of cells on the either side of the heart.
- Absorb colloidal particles
- e.g. *Apis* larva

Nitrogenous excretion

- Most aquatic insects excrete ammonia. Ammonia is highly toxic but soluble in water. Only insects that have access to lots of water can excrete ammonia.
- Terrestrial forms or those species that are water-limited excrete uric acid. Uric acid is insoluble in water and relatively non-toxic. The ability to form uric acid is often cited as one of the reasons members of the Arthropoda have been so successful on land relative to other invertebrates.
- Many terrestrial insects tend to lose water via evaporation while exchanging gases, while some plant sap feeders ingest so much water that they are challenged with excreting excess water and concentrating nutrients.

Circulatory System

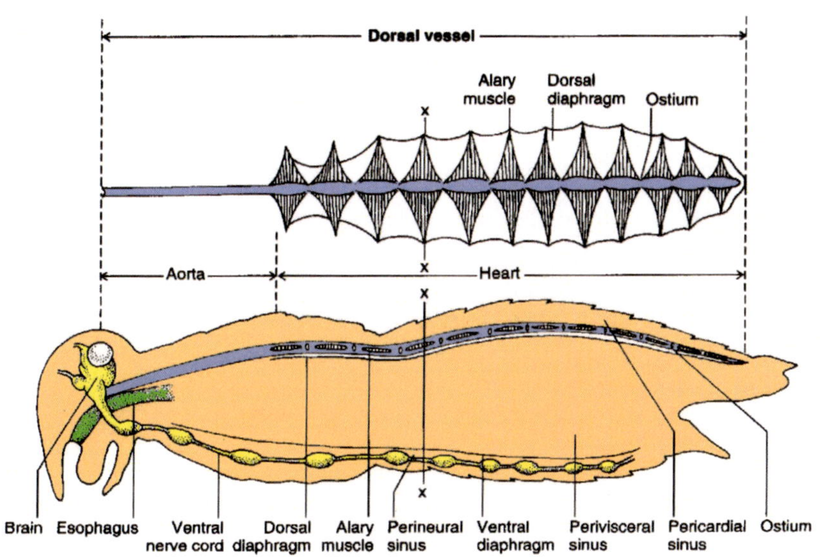

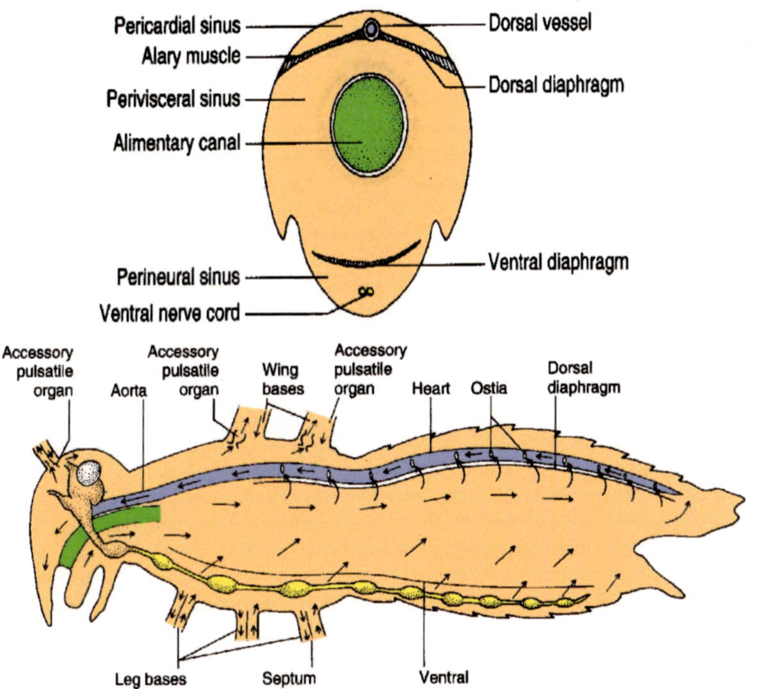

Open circulatory system

- Insects have an **open circulatory system** in which blood (usually called **haemolymph**) flowing freely within body cavities (Haemocoele) where it makes direct contact with all internal tissues and organs.

Haemocoele

- To facilitate circulation of haemolymph, the body cavity is divided into three compartments (called **blood sinuses**) by two thin sheets of muscle and/or membrane known as the dorsal and ventral diaphragms. The dorsal diaphragm is formed by alary muscles of the heart and related structures; it separates the pericardial sinus from the perivisceral sinus. The ventral diaphragm usually covers the nerve cord; it separates the **perivisceral sinus** from the **perineural sinus**.

Dorsal Blood vessel (DBV)

- It consists of an anterior aorta (located in the thorax) and a posterior series of hearts (located in the abdomen).
- Contractions move posterior to anterior and carry blood from the abdomen toward the head.

Heart

- It is a chambered tube lodged in sinus dorsally known as Dorsal or pericardial sinus cut off by a fibro-muscular septum called as Dorsal diaphragm.
- It is divided segmentally into chambers that are separated by valves (ostia) to ensure one-way flow of haemolymph.
- The heart is chambered; cockroach has 13 chambered heart, honeybee has 4 chambered heart; housefly has 3 chambered heart.
- Histologically it is composed of a single layer of cardiac cells carrying large nuclei. The cytoplasm contains within it striated muscle fibrillae disposed of in the periphery, and the cells have outer and inner sheaths probably derived from sarcolemma.
- Arising from the tergum and situated fanwise, a pair of **alary muscles** is attached laterally to the walls of each chamber.
- Peristaltic contractions of the muscles force the hemolymph forward from chamber to chamber. During each diastolic phase (relaxation), the ostia open to allow inflow of hemolymph from the body cavity.

- The heart's contraction rate varies considerably from species to species -- typically in the range of 30 to 200 beats per minute. The rate tends to fall as ambient temperature drops and rise as temperature (or the insect's level of activity) increases.

Functions of the heart

- transport of nutrients around the body
- movement of limbs, mouthparts, antennae
- molting -- by increasing pressure in certain parts of the body
- protection (immune response)
- thermoregulation

Aorta

- It is a simple tube (called the aorta) which lacks valves or musculature and continues forward to the head and empties near the brain.
- In odonata nymphs and *Dytiscus* aorta gives off diverticula in the thorax called as aortic ampullae
- Aorta is non-pulsatile but is pulsatile is *Dytiscus* and perform rhythmic contractions independent of heart.
- Histologically, the aorta is similar to heart.
- Haemolymph bathes the organs and muscles of the head as it emerges from the aorta, and then haphazardly percolates back over the alimentary canal and through the body until it reaches the abdomen and re-enters the heart.

Accessory Pulsatile organs

- In some insects, **pulsatile organs** are located near the base of the antennae, wings or legs.
- These muscular "pumps" do not usually contract on a regular basis, but they act in conjunction with certain body movements to force haemolymph out into the extremities.

Mechanism of circulation

- The circulation is chiefly affected by the pulsation of the heart. The heart rhythmically contracts from behind forwards. The contraction is simply myogenic but the rate and rhythm are controlled by nervous system.

- During diastole, heart expands; a negative pressure forces the blood from the pericardial sinus to enter the heart through the ostia.
- During systole a positive pressure is set up and the ostia get closed and blood is driven forwards to the chamber in front.
- The rate of heartbeat is variable; in cockroach it is 49/min; in *Hippobasca* 120/min.
- Heart beat vary with instar; highest when insect become active and lowest when insect enter diapauses (hibernation/aestivation) or in pupal period.

Blood/Haemolymph

Composition

- **Plasma (90%)**
    - A watery fluid -- usually clear, but sometimes greenish or yellowish in colour. Compared to vertebrate blood, it contains relatively high concentrations of amino acids, proteins, sugars, and inorganic ions (Chloride, Sodium, Magnesium). Organic constituents includes trehalose, glucose and fructose etc.
    - Pigments occur in plasma, like haemoglobin in *Chironomus* . The blood may be coloured due to pigments like :
    - Green insectoverdin : Lepidoptera and locusts
    - Blue mesobiliverdin : locusts
    - Purpulish red protaphin : aphids
    - Carotene, xanthophylls and flavins : *Bombyx* larva
    - Overwintering insects often sequester enough ribulose, trehalose, or glycerol in the plasma to prevent it from freezing during the coldest winters.
    - The blood is slightly acidic with pH 6-7.
- **Cellular (haemocytes)**
    - The remaining 10% of haemolymph volume is made up of various cell types (collectively known as **haemocytes**); they are involved in the clotting reaction, phagocytosis, and/or encapsulation of foreign bodies.
    - Blood cells forming (Haemopoietic) organs present in caterpillars are thoracic in position, capsulated or non-capsulated.
    - **Haemopoietic organs do not exists in adult insects.**
    - **Prohaemocytes** are rounded cells bearing large nuclei and basophilic cytoplasm. These cells divide and give rise to :

i. Plasmatocytes

- Most abundant, variable form, basophilic cytoplasm and phagocytic function.

ii. Granulocytes

- Cytoplasmic acidophilic granules. Cystocytes or coagulocytes are modified granulocytes

iii. Oenocytes

- Large basophilic cells

iv. Spherule cells

- Spheriodal, acidophilic cells.

v. Adipohaemocytes/spheroidocytes

- Contains fat droplets and other inclusions.
- The density of insect haemocytes can fluctuate from less than 25,000 to more than 100,000 per cubic millimeter, in general the haemocyte number in insects is 20,000- 40,000.mm^3.
- With the exception of a few aquatic midges, insect hemolymph does NOT contain hemoglobin (or red blood cells).
- Oxygen is delivered by the tracheal system, not the circulatory system.

Functions

- Does NOT carry O_2
- Transports nutrients, salts, hormones and metabolic waste products
- Seals off wounds by clotting
- Encapsulates and destroys parasites
- Hydrostatic pressure for molting and aids body movement
- Thermal regulation
- Serves as a lubricant for the movement of internal structures relative to one another.
- It is a hydraulic medium for applying pressure for molting, eversible glands are extruded via pressure changes, and some muscular contraction is opposed by hydrostatic pressure within the hemocoel.
- Haemolymph transports various substances from one tissue to another.

- Specialized cells that phagocytosize or encapsulate foreign particles are found in the hemolymph and are very important in the "immune" system of insects.
- Aids in thermoregulation

Fat bodies

- Mesodermal in origin in the form of loose mass of cells project freely in the haemocoele and are penetrated by a dense network of fine tracheole.
- Fat body is made up of Trophocytes.
- Serve as store house of food material.

Function

- Synthesis and secretion of :
  a. Larval-specific storage proteins (e.g. calliphorin)
  b. Vitellogenin (female-specific haemolymph protein)
  c. Lipoprotein and lipophorin (transport molecules)
  d. Juvenile hormone carrier proteins
  e. Juvenile hormone esterases
  f. Hemoglobin in Chironomus larvae (molecule is considered a storage protein)
  g. Diapause proteins
  h. Production of mixed-function oxidases (MFOs)

Respiratory System

All insects are aerobic organisms -- they must obtain oxygen (O_2) from their environment in order to survive. They use the same metabolic reactions as other animals (glycolysis, Kreb's cycle, and the electron transport system) to convert nutrients (e.g. sugars) into the chemical bond energy of ATP. During the final step of this process, oxygen atoms react with hydrogen ions to produce water, releasing energy that is captured in a phosphate bond of ATP.

The respiratory system is responsible for delivering sufficient oxygen to all cells of the body and for removing carbon dioxide (CO_2) that is produced as a waste product of cellular respiration. The respiratory system of insects (and many other arthropods) is separate from the circulatory system. It is a complex network of tubes (called a **tracheal system**) that delivers oxygen-containing air to every cell of the body.

The tracheal system is a cuticle lined system of tubes that extend inward from openings along the sides of the body called spiracles. Trachea is strengthened by **taenidia** – ring like annulations along the length that give the trachea resistance to compression. The smallest branches of the tracheal system are called tracheoles and they are on the order of 0.1 to 0.2 um in diameter.

Most Collembola, many Protura, and certain endoparasitic wasp larvae lack a tracheal system; gas exchange occurs via the integument.

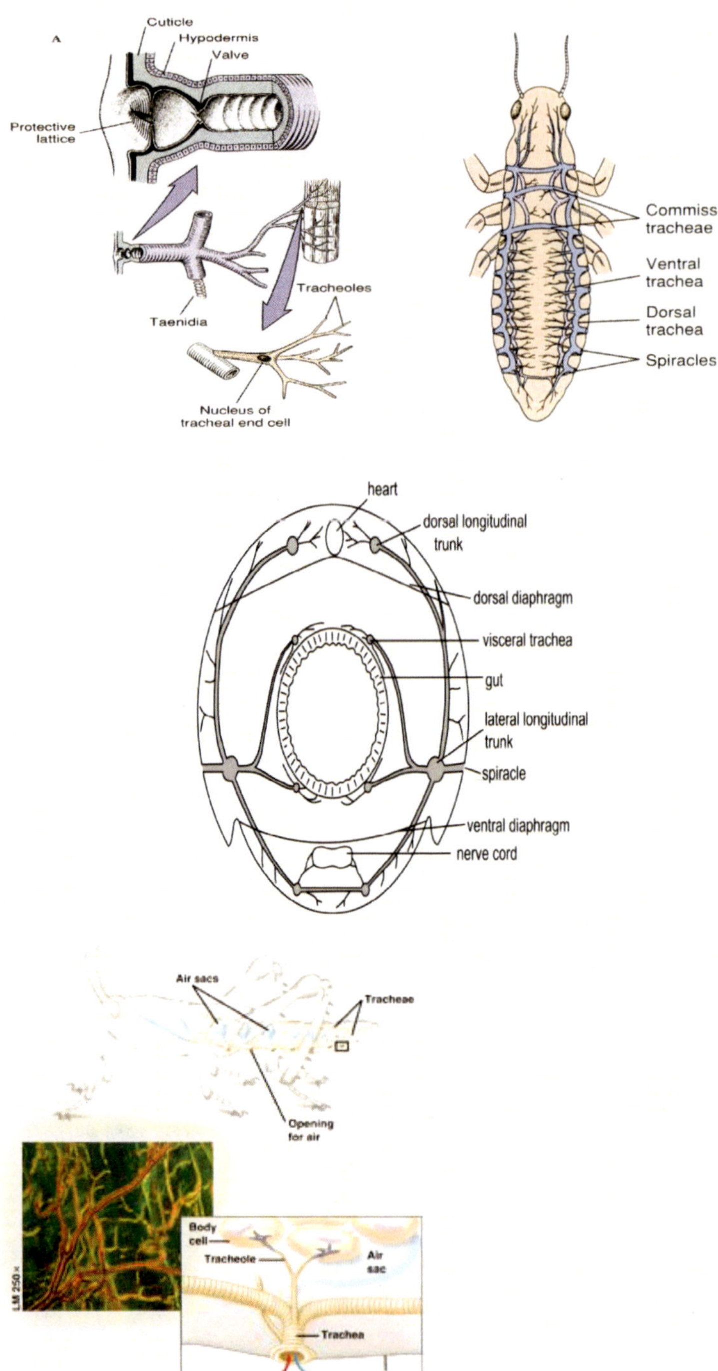
A
Cuticle
Hypodermis
Valve
Protective lattice
Tracheoles
Taenidia
Nucleus of tracheal end cell
Commissural tracheae
Ventral trachea
Dorsal trachea
Spiracles
heart
dorsal longitudinal trunk
dorsal diaphragm
visceral trachea
gut
lateral longitudinal trunk
spiracle
ventral diaphragm
nerve cord
Air sacs
Tracheae
Opening for air
LM 250×
Body cell
Tracheole
Air sac
Trachea
Body wall
O_2
CO_2

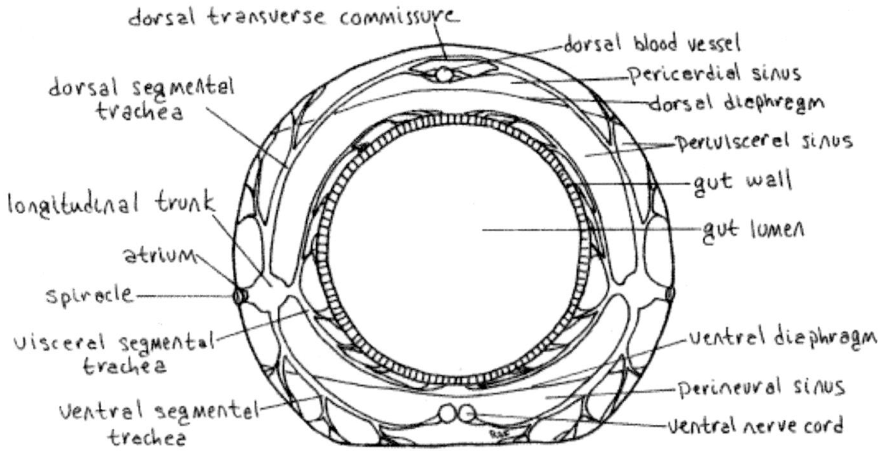

Spiracles

- Air enters the insect's body through valve-like openings in the exoskeleton. These openings (called **spiracles**) are located laterally along the thorax and abdomen of most insects -- usually one pair of spiracles per body segment.
- Air flow is regulated by small muscles that operate one or two flap-like valves within each spiracle -- contracting to close the spiracle, or relaxing to open it.
- Simple spiracles are merely openings to the outside.
- **Atriate spiracles** have mechanisms that allow the insect to close the opening.
  - Prevents water loss
  - Prevents the entry of pathogens and parasites
- The spiracles leads in to short spiracular trachea subsequently dividing in to:

  a. Dorsal tracheal trunk supplying the dorsal muscles and dorsal muscles

  b. Visceral trachea supplying the digestive system, fat bodies and gonads

  c. Ventral trachea supplying ventral musculature, nerve cord, legs etc.

- The main longitudinal trunk is formed as a fusion of the dorsal, ventral and visceral trachea.

- In most insects the tracheae are all linked through a series of longitudinal pipes called trunks and many smaller connections. Dorsal Longitudinal Trunk near the top, or back, of the insect's body; Lateral Longitudinal Trunk running along the sides just in from the spiracles; Ventral Longitudinal Trunk running along the belly of the insect
- Tracheal commissures or transverse connectives and ventral commissures connecting the tracheal system of each side.
- The number of spiracles an insect has is variable between species, however they always come in pairs, one on each side of the body, and usually one per segment.
- Some of the **Diplura have eleven pairs, with four pairs on the thorax,** but in most of the ancient forms of insects, such as **Dragonflies and Grasshoppers there are two thoracic and eight abdominal spiracles.**
- However in most of the remaining insects there are less; so that **Hoverflies, Syrphidae, have only two pairs, both of which are on the thorax and none on the abdomen** while many **Mosquito larvae and aquatic Beetle larvae have only one abdominal pair of spiracles**.
- Many insects have valves that allow them to close their spiracles, thus preventing water loss.
- In some of the **Collembola each spiracle produces a tree branch, or tree root, of tracheae that are separate from those of other spiracles**.

Tracheae

- Each tracheal tube develops as an invagination of the ectoderm during embryonic development.
- To prevent its collapse under pressure, a thin, reinforcing "wire" of cuticle (the **taenidia**) winds spirally through the membranous wall. This design gives tracheal tubes the ability to flex and stretch without developing kinks that might restrict air flow.
- After passing through a spiracle, air enters a longitudinal tracheal trunk tracheal trunk, eventually diffusing throughout a complex, branching network of tracheal tubes that subdivides into smaller and smaller diameters and reaches every part of the body. At the end of each tracheal branch, a special cell (the **tracheole**) provides a thin, moist interface for the exchange of gasses between atmospheric air and a living cell.
- Oxygen in the tracheal tube first dissolves in the liquid of the tracheole and then diffuses into the cytoplasm of an adjacent cell. At the same time, carbon dioxide, produced as a waste product of cellular respiration, diffuses out of the cell and, eventually, out of the body through the tracheal system.

- They are also lined with intima that is shed during each molt.
- Chitin is lacking in the smaller tracheae.

Air sacs

- Air sacs are found mainly in **flying insects.**
- Many potential functions, all rather speculative, including increasing the volume of air in the body for exchange, lowering the specific gravity for flight, and providing room for the growth of internal organs.
- The **absence of taenidia** in certain parts of the tracheal system allows the formation of collapsible air sacs, balloon-like structures that may store a reserve of air.
- In **dry terrestrial environments**, this temporary air supply allows an insect to conserve water by closing its spiracles during periods of high evaporative stress.
- Aquatic insects consume the stored air while under water or use it to regulate buoyancy. During a molt, air sacs fill and enlarge as the insect breaks free of the old exoskeleton and expands a new one. Between molts, the air sacs provide room for new growth -- shrinking in volume as they are compressed by expansion of internal organs.

Tracheoles

- The smallest branches of the tracheae are the tracheoles.
- Gaseous exchange occurs here.
- Provide for a tremendous amount of surface area (**there are 1.5 million tracheoles in a large silkworm larva**).

Types of tracheal systems

- Based on the location and functioning of the spiracles, the treacheal systems are classified in to **4 types** :

a. Polypneustic type

- At least 8 functional spiracles on each side
    - **Holopneustic type**
        - 10 spiracles functional; 1 mesothoracic + 1 metathoracic + 8 abdominal spiracles are functional. Eg. *Bibionid* larvae

- **Peripneustic type**
  - 9 spiracles functional; 1 mesothoracic + 1 metathoracic + 7 abdominal spiracles are functional. Eg. Ceccidomyid larvae
- **Hemipneustic type**
  - 8 spiracles functional; 1 mesothoracic + 7 abdominal spiracles are functional. Eg. Mycetophilid larvae

b. Oligopneustic type

- One or two spiracles are functional on each side
  - **Amphipneustic type**
    - 2 spiracles functional; 1 mesothoracic + 1 post abdominal spiracles are functional. Eg. Psychodid larvae
  - **Metapneustic type**
    - 1 spiracle functional; 1 post abdominal spiracle is functional. Eg. Culicid larvae
  - **Propneustic type**
    - 1 spiracle functional; 1 mesothoracic spiracle is functional. Eg. Dipterous larvae

c. Apneustic type

- None of the spiracles being functional, air entering the tracheal system by diffusion through the general body surface or gills.
- Closed systems are found in many aquatic insects and the larvae of endoparasitic species. Closed systems lack spiracles and gaseous exchange between the tracheal system and the environment occurs directly through the integument.

d. Hypopneustic type

- Without full complements of ten spiracles, but where one or more pairs disappear completely as **in thrips with two pairs of thoracic and two pairs of abdominal spircales, or only two pairs as in scale insects**.

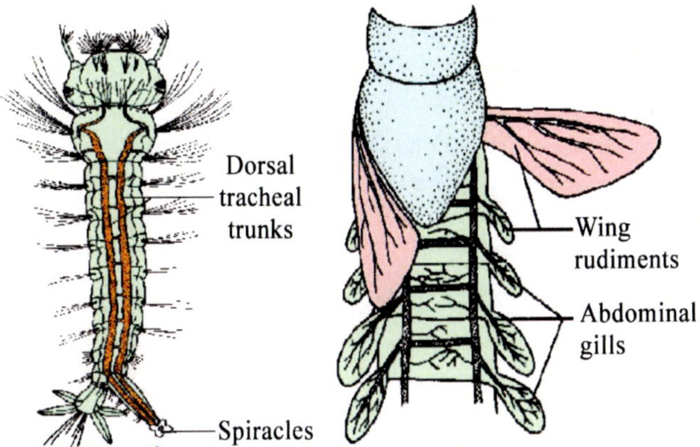

Mechanism of respiration

- The entry of air in to the tracheal system is controlled by the opening and closing of spiracles.
- The control of respiratory activities results from the diffusion control due to the opening and closing of spiracles is regulated by the nervous system.
- During inspiratory phase air is entered through spiracles and trachea. The main trunks and air sacs become filled with air. During expiratory phase air is compressed out. There is an increase in pressure during the compression of the abdomen and this helps in expansion of the smaller tubes.
- Compression and decompression drive the air deeper and deeper in to the tracheoles carrying the air directly to the cells of the tissues.\
- The amount of CO_2 produced in metabolism is generally lesser than the O_2 consumed and as its rate of diffusion through air is slower, CO_2 passes through the same route i.e. come out through the spiracles.
- **A stationery insect generally uses 0.6-3 litres of O_2/kh/hr but during flight it increases from 12 to 180 litres.**

Ventilation in insects

- **Passive ventilation**
  - The insect employs no pumping or other movements to aid the passage of gases into and out of the tracheae.
  - The insect may control the flow by opening and closing the spiracles.
  - Generally effective for smaller insects and those with well-ventilated air sacs.

- **Active ventilation**
  - The insect increases air movement by pumping movements of the abdomen, thorax, and even protraction and retraction of the head.
  - In many larger bodied and/or active insects requiring large volumes of gas for exchange, the movements of the body are coordinated with opening and closing of the spiracles to produce a unidirectional flow of gases through the body.
  - Air moves into the animal via the thoracic spiracles and out of the body via the abdominal spiracles.

Respiration in Aquatic Insects

Although water is a liquid, it usually contains a significant amount of **dissolved oxygen (4.9%)** plus small amounts of other gasses. Aquatic insects are equipped with a variety of adaptations that allow them to carry a supply of oxygen with them under water or to acquire it directly from their environment.

Hydrofuge hairs (e.g. Culicid larvae)

- Special hairs having a coating of non wettable wax on sides of spiracles enavling the larvae to stay in water surface and expose their spiracles to the air.
- **hydrofuge structure** functions in breaking the surface film of the water thereby exposing the spiracles to the atmosphere.
- Hydrofuge structures are usually made of "hairs" that are resistant to wetting causing the water near the surface to be repelled from the hydrofuge areas.

Tracheal gills

- Many larger bodied aquatic insects posses **tracheal gills**, integumental evaginations covered by a very thin cuticle and well supplied with tracheae and tracheoles.
- An extensive network nework of fine tracheal beneath the integument, either concentrated in some regions or distributed all over.
- E.g. tracheal gills of larvae of Plecoptera, Trichoptera and Ephemeroptera

Air stores

- Many aquatic Hemiptera and Coleoptera carry air stores in the form of bubbles.
- These bubbles are held in place by hydrofuge hairs and/or the shape of the body forms a storage area.

- Many insects carry stores of air and are able to replenish the oxygen without surfacing.
- Stores of oxygen act as a "**physical gill**" where the partial pressure differential of the oxygen between the gill and the water results in oxygen diffusing into the gill from its dissolved state. Nitrogen in the air store does not readily diffuse into the water and thus prevents the air store from collapsing.
- A **plastron** is a very thin layer of gas held firmly in place by tiny hydrofuge hairs.
- The spiracles open directly into this thin layer.
- Other insects tap directly into aquatic plants to obtain oxygen.

Spiracular gills

- Filamentous outgrowth of the ectoderm without any connection to the body cavity are developed.
- The Cuticular wall become excavated to form air spaces which become connected to the tracheal system.
- Any air filled space which is in contact with spairacles makes available its oxygen·to the tarcehal system thus functioning as temporary stores.
- E.g. Coleopotera and Diptera

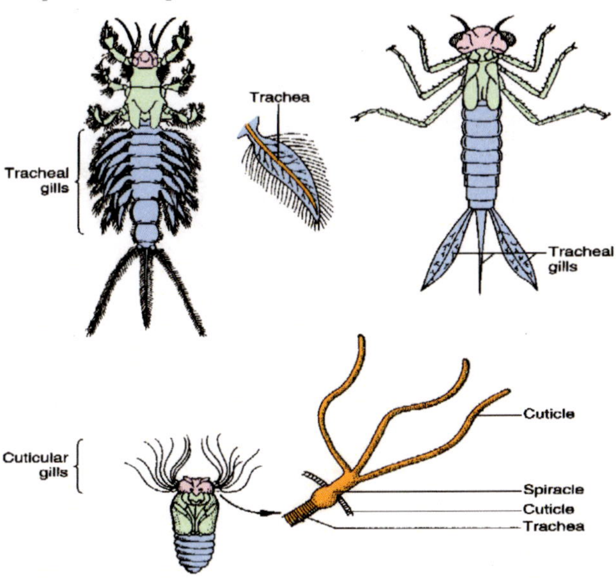

Cuticular Respiration

- Diffusion of gasses through a relatively thin integument that is permeable to oxygen (and carbon dioxide) sufficient to meet the metabolic demands of small, inactive insects -- especially those living in cold, fast-moving streams where there is plenty of dissolved oxygen.

Biological Gills

- A **biological gill** is an organ that allows dissolved oxygen from the water to pass (by diffusion) into an organism's body. In insects, gills are usually outgrowths of the tracheal system.
- They are covered by a thin layer of cuticle that is permeable to both oxygen and carbon dioxide.
- In mayflies and damselflies, the gills are leaf-like in shape and located on the sides or rear of the abdomen. Fanning movements of the gills keep them in contact with a constant supply of fresh water.
- Stoneflies and caddisflies have filamentous gills on the thorax or abdomen.
- Dragonflies differ from other aquatic insects by having internal gills associated with the rectum.

Biological gill

Breathing tubes

Air bubble

Breathing Tubes

- Although many aquatic insects live underwater, they get air straight from the surface through hollow breathing tubes (sometimes called **siphons**) that work on the same principle as a diver's snorkel.
- In mosquito larvae, for example, the siphon tube is an extension of the posterior spiracles. An opening at the end of the siphon is guarded by a ring of closely spaced hairs with a waterproof coating. At the air-water interface, these hairs break the surface tension of the water and maintain an

open airway. When the insect dives, water pressure pushes the hairs close together so they seal off the opening and keep water out. Water scorpions (Hemiptera: Nepidae) and rat-tailed maggots (larvae of a Syrphid fly) are two more examples of aquatic insects that have snorkel-like breathing tubes.

- Many aquatic plants maintain their bouyancy by storing oxygen (a waste product of photosynthesis) in special vacuoles. A few insects (*e.g.* larvae of *Mansonia* spp. mosquitoes) insert their breathing tubes into these air stores and obtain a rich supply of oxygen without ever swimming to the surface of the water.

Air Bubbles

- Some aquatic insects (diving beetles) carry a bubble of air with them whenever they dive beneath the water surface. This bubble may be held under the elytra (wing covers) or it may be trapped against the body by specialized hairs. The bubble usually covers one or more spiracles so the insect can "breathe" air from the bubble while submerged.
- An air bubble provides an insect with only a short-term supply of oxygen, but it also "collect" some of the oxygen molecules dissolved in the surrounding water. In effect, the bubble acts as a "physical gill" -- replenishing its supply of oxygen through the physics of passive diffusion.

Plastrons

- A **plastron** is a special array of rigid, closely-spaced hydrophobic hairs (setae) that create an "airspace" next to the body.
- Air trapped within a plastron operates as a physical gill (just like air in a bubble) but this airspace cannot shrink in volume because the fortress of setae prevents encroachment of surrounding water.
- When the insect consumes oxygen, it creates a partial pressure deficit inside the plastron. This deficit is "corrected" by dissolved oxygen that diffuses in from the water. As nitrogen gradually diffuses out of the bubble, it creates a similar partial pressure deficit. But there is very little dissolved nitrogen present in water (it has a lower solubility potential than oxygen), so some of the nitrogen's partial pressure deficit is "corrected" by oxygen.
- In effect, the plastron "trades" some of the nitrogen for oxygen -- keeping a constant volume of gas that may slowly become "enriched" with oxygen.
- The constant volume of a plastron's air supply eliminates the periodic need to surface and replenish the bubble. Insects that remain permanently submerged (ex. riffle beetles, family Elmidae) or lack the ability to reach

the surface (ex. eggs of floodwater mosquitoes) are likely to have plastrons. These structures are often visible underwater as thin, silvery films of air covering parts of the body surface.

Haemoglobin

- Haemoglobin is a respiratory pigment occurs only rarely in insects -- most notably in the larvae of certain midges (family Chironomidae) known as bloodworms, *Chironomus, Gasterophilus* and *Anisops*. These distinctive red "worms" usually live in the muddy depths of ponds or streams where dissolved oxygen may be in short supply.
- Under normal (aerobic) conditions, hemoglobin molecules in the blood bind and hold a reserve supply of oxygen. Whenever conditions become anaerobic, the oxygen is slowly released by the haemoglobin for use by the cells and tissues of the body. This back-up supply may only last a few minutes, but it's usually long enough for the insect to move into more oxygenated water.

Respiration in endo-parasites

- The exchange of gases takes place between the tissues fluid of the parasites and the host.
- Many parasites have caudal vesicles or filaments or processes which are respiratory in function.
- Sometimes the front of the egg shell may be modified in to Aeroscopic plate.
- In parasitic hymenopetran insects, the pedicel of the egg is long and protrudes externally through the body wall of the host and functions as a sort of respiratory organ.

The Insect Nervous System

Introduction

Insects have well-developed senses relative to many other invertebrates

Function of Nervous system

- Detect any changes - Inside the body; Outside the body
- Centralize information
- Collect information and Analyze information
- Respond to changes : Express as behaviour;
- Secrete hormones

Common insect neurotransmitters include

- Acetylcholine
- glutamic acid (primarily in muscles)
- GABA (inhibits further depolarization)
- some pesticides act by inhibiting functioning of the nervous system, esp. neurotransmitters
- organophosphates (malathion, and others) disrupt activity of acetylcholinesterase

A fundamental difference between invertebrate and vertebrate nervous systems is the number of cells

- molluscs: few thousand neurons
- insects: 500,000 neurons
- vertebrate: 10 billion neurons

A mere 44 millisecs from detection to initiation of evasive response!

The neuron

- The basic functional unit of the nervous system is the nerve cell or neuron.
- The neuron consists of a cell body, one or more axons, and dendrites.
- Three general types of neurons
  - Unipolar neurons have a single stalk from the cell body that connects with the axon.
  - In bipolar neurons, the cell body bears an axon and a single, branched or un-branched dendrite.
  - Multipolar neurons have an axon and several branched dendrites.
- Neurons 'communicate' with one another and with other cells either electrically or chemically via neurotransmitters through small spaces called synapses.
- Actions may be either excitatory or inhibitory.
- Areas where neurons are concentrated are called ganglia.
- Nerves are bundles of axons.

Four Types of Insect Nerves

- *Sensory nerves*: pick up stimulus and generate an action potential
- *Motor nerves*: effect a response, e.g., gland releases or muscle movement
- *Interneurons*: nerve to nerve communication
    - comprise the body of the central nervous system
    - responsible for setting up endogenous rhythms (of activity, hormone release, etc.)
- *Neuroendocrine cells*: secrete hormones

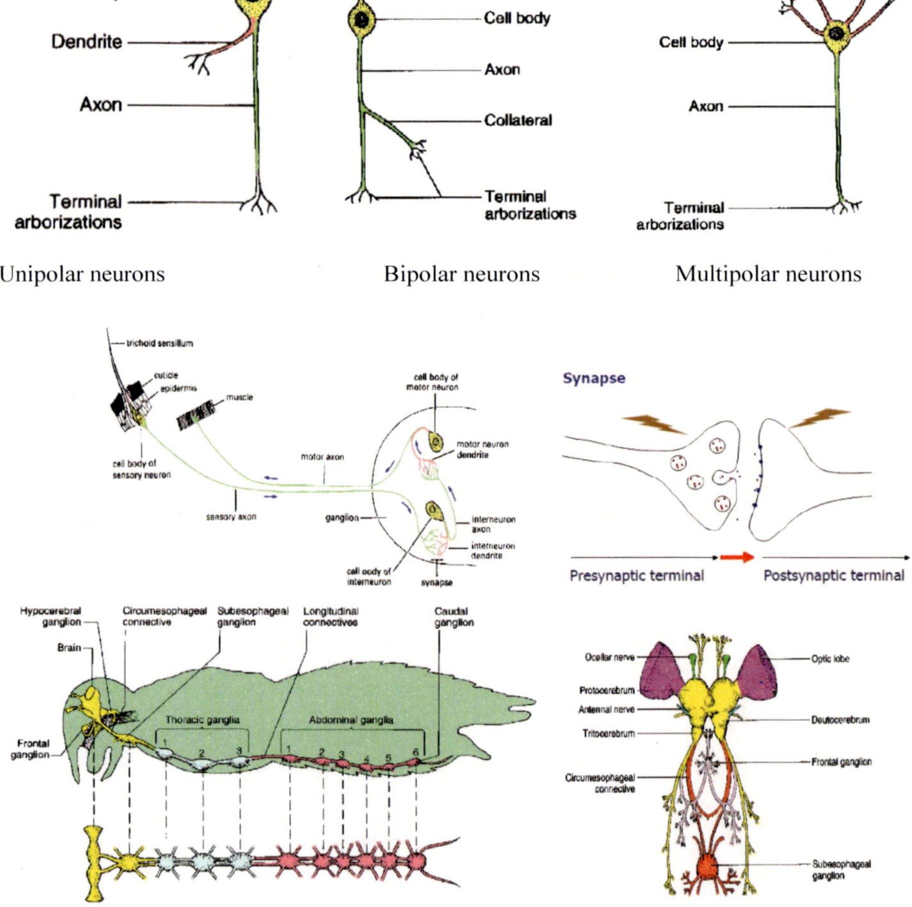

Central Nervous System

Like most other arthropods, insects have a relatively simple central nervous system with a dorsal brain linked to a ventral nerve cord that consists of paired **segmental ganglia** running along the ventral midline of the thorax and abdomen. Ganglia within each segment are linked to one another by a short medial nerve (**commissures**) and also joined by inter-segmental **connectives** to ganglia in adjacent body segments. In general, the central nervous system is rather ladder-like in appearance: **commissures are the rungs of the ladder and inter-segmental connectives are the rails**. In more "advanced" insect orders there is a tendency for individual ganglia to combine (both laterally and longitudinally) into larger ganglia that serve multiple body segments.

Brain

An insect's **brain** is a complex of six fused ganglia (three pairs) located dorsally within the head capsule. Each part of the brain controls (innervates) a limited spectrum of activities in the insect's body.

a. The central nervous system is composed of a double chain of ganglia joined by longitudinal connectives.

b. The anterior ganglion is the brain. The brain connects to the ventral chain of ganglia via two connectives that travel around the pharynx. The brain connects to the eyes, ocelli, and antennae.

c. The subesophageal ganglion is highly complex and innervates the sense organs and muscles associated the mouthparts, the salivary glands, and the neck region. In many insects the subesophageal ganglion is also the primary excitatory or inhibitory influence on motor activity of the whole insect.

d. The frontal ganglion connects the brain to the stomatogastric subsystem.

e. The hypocerebral ganglion is associated with two endocrine glands one of which is the corpus allatum that produces JH (juvenile hormone).

f. The thoracic ganglia contain the sensory and motor centers for their respective segments. In some insects these three ganglia are fused into one.

Protocerebrum

- The first pair of ganglia are largely associated with vision;
- Innervates eyes, optic lobes, labrum, and median ocellus
- coordination of complex behaviors and most learning
- Neuroendrocrine control of molting (PTTH secreted from protocerebrum)

- *Mushroom bodies* located dorsomedially between optic lobes
- Important in smell and may be involved in learning

Deutocerebrum

- Process sensory information collected by the antennae,
- Innervates antenna, olfactory lobes, and lateral ocelli

Tritocerebrum

- Innervate the labrum and integrate sensory inputs from proto- and deutocerebrums.
- Link the brain with the rest of the ventral nerve cord and the stomodaeal nervous system
- Innervates and control ventral nerve cord

Subesophageal Ganglion

- Located ventrally in the head capsule (just below the brain and oesophagus) is another complex of fused ganglia (jointly called the sub-oesophageal ganglion).
- innervates mouthparts (fusion of ganglia from segments 4-6 from the embryo, i.e., the mandible, maxilla, and labium)
  - mandibular ganglion -- sensory nerves from the mandibles
  - maxillary ganglion -- sensory nerves from the maxillae
  - labial ganglion -- sensory nerves from the labium
- A pair of circumesophageal connectives loops around the digestive system to link the brain and subesophageal complex together.

Ventral Nerve Cord

Thoracic ganglia:

- Three pairs of thoracic ganglia (sometimes fused) control locomotion by innervating the legs and wings.
- Thoracic muscles and sensory receptors are also associated with these ganglia.

Abdominal ganglia:

- Genitalia, heart, excretion, etc.

- Control movements of abdominal muscles.
- Spiracles in both the thorax and abdomen are controlled by a pair of lateral nerves that arise from each segmental ganglion (or by a median ventral nerve that branches to each side).
- A pair of terminal abdominal ganglia (usually fused to form a large **caudal ganglion**) innervate the anus, internal and external genitalia, and sensory receptors (such as cerci) located on the insect's back end.

Visceral/Stomodaeal or stomatogastric nervous system

- Nerves associated with the brain, salivary glands, and the foregut are the stomatogastric subsystem. These include the frontal ganglion and the hypocerebral ganglion.
- The nerves associated with the ventral nerve cord are the ventral viscera subsystem.
- The caudal visceral subsystem is associated with the posterior segments of the abdomen (the caudal region) including the reproductive system.
- A pair of frontal nerves arising near the base of the tritocerebrum link Oesophagus the brain with a frontal ganglion (unpaired) on the anterior wall of the esophagus. This ganglion innervates the pharynx and muscles associated with swallowing. A recurrent nerve along the anterio-dorsal surface of the foregut connects the frontal ganglion with a hypocerebral ganglion that innervates the heart, corpora cardiaca, and portions of the foregut. Gastric nerves arising from the hypocerebral ganglion run posteriorly to ingluvial ganglia (paired) in the abdomen that innervate the hind gut.

Peripheral nervous system

- All of the nerves with synapses to the central and the visceral nervous systems
- These nerves are associated with sensory structures.

Hormones produced by Brain

Prothoracicotropic Hormone (PTTH)	: Stimulates prothoracic gland to synthesis ecdysone. Ecdysone is a hormone that initiates ecdysis
Eclosion hormone (EH)	: Control muscular movement in ecdysis to become adults

Brain or ventral nerve cord

Bursicon	: Initiates sclerotization after ecdysis
Neurohemal organs	: Organs that releases hormones
Prothoracic gland	: Synthesize and release ecdysone/ Moulting hormone Store and release PTTH
Corpora allata (CA)	: Release Juvenile Hormone (JH) Preserve larval characteristics during the molts Inhibit metamorphosis Release bursicon

Endocrine System

Neurosceretory cells (NSC): Includes

a. Chiefly lying in the mid region of the brain and composed of two or more distinct types of NSC.

b. Lateral NSC are also located in the brain

c. NSC of the ganglia of the CNS

- Functions
  - The brain produces Prothoracotropic hormone (Brain Hormone) which stimulates the Thoracic glands to produce the moulting hormone (MH).
  - In *Rhodnius,* the NSC produce a Diuretic Hormone inducing diuresis.
  - In Blowflies, NSC produces Bursicon facilitating darkening and hardening of cuticle of adults.

Corpora Cardiaca (CC)

- A pair of ganglia like glands lying in the close association with the axons of the NSC of the brain and situated lateral to aorta.
- Neurosecretions from the NSC of the brain are stored temporarily and periodically released in the blood.

Corpora allata (CA)

- Pair of ganglia like glands lying ventral to the anterior end of the aorta.
- In larvae of Diptera, a ring shaped gland (**Weismann's ring**) comprises a complex organ of endocrine centres- a ventral CA, a dorsal CC and two lateral tracheal glands which correspond to the thoracic gland.

- The CA produces Juvenile Hormone (JH)/Neotenin which maintains larval character.

Thoracic glands/Ventral glands

- A pair of ganglia like glands lying in the close association with the axons of the NSC of the brain and situated lateral to aorta.
- Lyonet (1762) firts describe these glands in caterpillars.
- It produces Moulting Hormone (**MH**)/Ecdysone, first isolated by Butenandt and Karlson (1954)

Pheromones

Semiochemicals are chemical substances that mediate communication between organisms. Semiochemicals maybe classified into Pheromones (intraspecific semiochemicals) and Allelochemics (interspecific semiochemicals).

Pheromones are chemicals secreted into the external environment by an animal which elicit a specific reaction in a receiving individual of the same species. Pheromones are volatile in nature and they aid in communication among insects.

Pheromones are exocrine in origin (i.e. secreted outside the body). Hence they were earlier called as ectohormones. In 1959, German chemists Karlson and Butenandt isolated and identified the first pheromone, a sex attractant from silkworm moths. They coined the term pheromone. Since this first report, hundreds of pheromones have been identified in many organisms. The advancement made in analytical chemistry aided pheromone research.

Based on the responses elicited pheromones can be classified into 2 groups

a) **Primer pheromones:** They trigger off a chain of physiological changes in the recipient without any immediate change in the behaviour. They act through gustatory (taste) sensilla. (eg.) Caste determination and reproduction in social insects like ants, bees, wasps, and termites are mediated by primer pheromones. These pheromones are not of much practical value in IPM.

b) **Releaser pheromones:** These pheromones produce an immediate change in the behaviour of the recipient. Releaser pheromones may be further subdivided based on their biological activity into

   1) Sex pheromones

   2) Aggregation pheromones : Bark beetles

   3) Alarm pheromones : Honeybees,

   4) Trail pheromones : Ants

Sex pheromones

- Butenandt and his coworkers in 1959 isolated 12mg of pheromone from the abdomen of half a million virgin females of silkworm. They named the pheromone as Bombykol. The chemical name is 10,12 – hexadeca dienol. It is a primary alcohol.
- The following are some of the female sex pheromones identified in insects

Sl.No.	Name of the Insect	Pheromone
1.	Silkworm, *Bombyx mori*	Bombykol
2.	Gypsy moth, *Porthesia dispar*	Gyplure, disparlure
3.	Pink bollworm ,*Pectinophora gossypiella*	Gossyplure
4.	Cabbage looper, *Trichoplusia ni*	Looplure
5.	Tobacco cutworm, *Spodoptera litura*	Spodolure, litlure
6.	Gram pod borer, *Helicoverpa armigera*	Helilure
7.	Honey bee queen, *Apis sp.*	Queen's substance

Glandular System

Glandular system is otherwise called as secretary system and is divided in to two major groups based on the presence or absence of ducts.

A. **Exocrine glands** (glands with duct)

1. **Salivary glands:** Salivary glands are modified labial glands which secrete saliva and open beneath hypopharynx.
2. **Mandibular glands:** Secrete saliva in caterpillars when salivary glands are modified into silk glands. In queen bee it secretes queen substance.
3. **Maxillary glands:** Secretions are useful to lubricate mouth parts.
4. **Pharyngeal glands:** Secrete bee milk or royal jelly in nurse bee.
5. **Frontal glands:** Secrete sticky defensive fluid in nasute termites.
6. **Setal glands:** Glandular seta (Scoli) secrete irritant fluid in hairy/slug caterpillar.
7. **Tenant hairs:** Secrete sticky fluid found in pulvilli of legs and helps in ceiling walking in house flies.
8. **Moulting glands:** Modified glandular epidermal cells, secrete moulting fluid necessary for moulting.
9. **Stink glands (Repugnatorial glands):** Secrete bad smelling substance. e.g. Stink bugs, bed bugs.
10. **Osmeteria (Forked gland):** Eversible gland in the thorax of papilionid larva with defense function. e.g. Citrus butterfly larva.

11. **Androconia (Scented scales):** Secretions of glandular scales of male pierid butterflies to attract the opposite sex.
12. **Pheromone glands:** Found in abdominal terminalia of one sex and its secretions are released outside to attract opposite sex of the same species.
13. **Wax glands:** Dermal glands producing wax in bees and mealy bugs.
14. **Sting glands:** Modified accessory glands secreting venom in worker bees and wasps.
15. **Lac glands:** Dermal glands secreting resinous substances in lac insect.
16. **Milk glands:** Modified accessory gland nourishing larva developing in uterus. e.g. Sheep ked.

B. **Endocrine glands** (glands without duct)

1. **Neurosecretory cells**: A pair of median neuro-secretory cells and lateral neurosecretory cells are present. The axons of these neurosecretory cells form two pairs of nervi corpora cardiaci ending in carpora cardiaca. This structure influence the functioning of other endocrine glands.
2. **Corpora cardiaca**: It consist of paired bodies fused in middle and have both nervous tissues and glandular tissues. It acts as a conventional storage and release organ for neurosecretory cells. It controls heart beat and regulate trehalose level in haemolymph.
3. **Corpora allata**: It is a paired gland attached to corpora cardiaca and secretes juvenile hormone (**JH**) there by inhibit metamorphosis. It is needed for egg maturation and functioning of male accessory glands. Practically JH analogues interfere with insect development. **Precocene** is an anti JH which induce precocious metamorphosis and death in insects.
4. **Prothoracic glands**: Paired gland present in ventrolateral part of prothorax of larva and is degenerated in adults. It secretes the moulting hormone **ecdysone**. Neurosecretory cells activate prothoracic glands to secrete ecdysone.
5. **Weismann's ring**: Formed by the fusion of carpora cardiaca, carpora alleta, prothoracic glands and hypocerebral ganglion to secrete puparium hardening hormone. Present in maggots of Dipteran flies.

Semio Chemicals

Definition

Chemicals that deliver behavioural messages which act either interspecially or intraspecifically.

Interspecific semiochemicals

1. **Allomone** - Interspecific semiochemical that favours the produces E.g. Repellents, Deterrents (feeding and ovipositional)
2. **Kairomone** - Interspecific semiochemical that favours the receiver E.g. attractants ' Food love"
3. **Synamone** - Interspecific semiochemical that favours both the producer and receiver E.g Plant odours attracting natural enemies of pests.
4. **Apneumone**- Pheromone like substance produced by the non living material that benefits a reciepient but detrimental to different species associated with the non living material

Intraspecific semiochemicals

1. **Pheromone-** Semiochemical used for intraspecific communication which is an exocrine secretion that causes specific reaction in the receiving individuals of the same species.
2. **Sex pheromone-** Female produce to attract males E.g. Bombykol (*Bombyx mori*) Gyplure (gypsy moth) and Gossyplure (Pink boll worm) (In American boll weevil males produce)
3. **Alarm pheromone-** Semiochemicals used to warn other fellow individuals from mandibular glands or anal glands. E.g. honey bees (E) B. Farnesene aphids.
4. **Trailmarking pheromone-** Semiochemicals used in route perception. Eg. Ants, termites.
5. **Aggregation pheromone-** Semiochemicals which attract other fellow members to a particular spot. E.g. Ferrolure of red palm weevil.

Reproductive System

Why Reproduce

- *Driving force behind evolution after survival is Reproduction.*
- *Surviving alone without leaving progeny is useless and costly*
- *To leave off springs or else that is the end of that particular spp. continuation of spp*

- *Sexual reproduction provides a mechanism for shuffling and recombining genetic information from two parents to create new genotypes that can be tested in the fire of natural selection. Only phenotypes that withstand the "heat" can participate in the next round of reproduction*

Why we need to know reproduction

Applied in different field like

- Sterile male release technique or SIT (sterile insect technique)
- IPM strategies-Knowing how long it takes female to develop eggs and then to lay them
- Forensics

External Fertilization

- Primitive arthropods (**Myriapods, Silverfish, bristletails and Collembola**) living in the water, their sperm could simply swim from the male's body to the female's body where fertilization could occur
- The large numbers of sperms are encapsulated within a water-tight lipoprotein shell secreted by the male's accessory glands. These "packages" of sperm are known as **spermatophores**.
- In myriapods and primitive hexapods (*e.g.* Collembola), males leave spermatophores on the ground where they may be found and picked up by a passing female. have more elaborate courtship activities in which the male leads his mate to a freshly deposited spermatophore
- Most Apterygota:
    - Indirect sperm transfer
    - Requires spermatophore and generally humid conditions

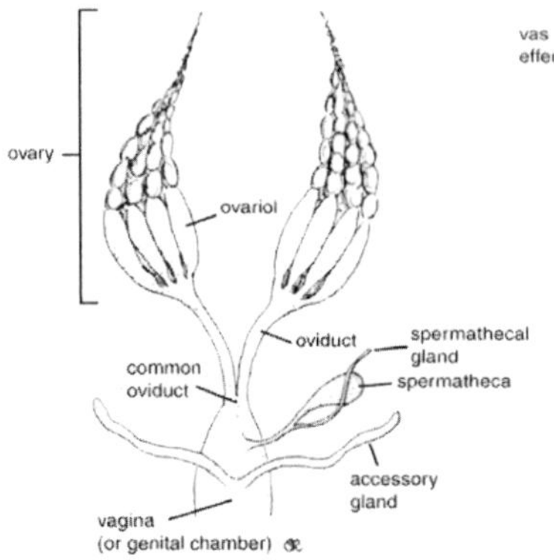

ovary
ovariol
oviduct
spermathecal gland
spermatheca
common oviduct
accessory gland
vagina
(or genital chamber)

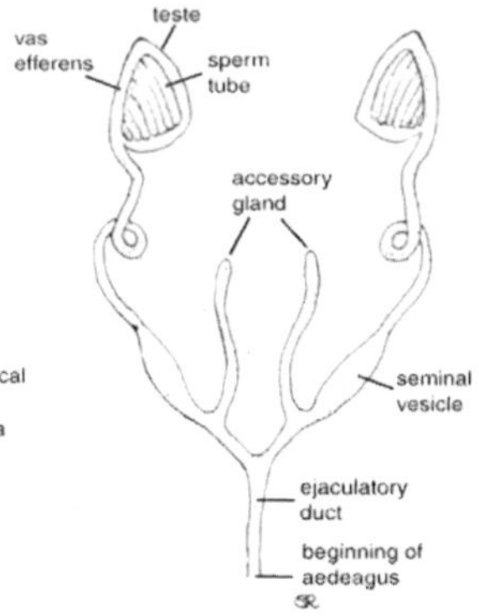

teste
vas efferens
sperm tube
accessory gland
seminal vesicle
ejaculatory duct
beginning of aedeagus

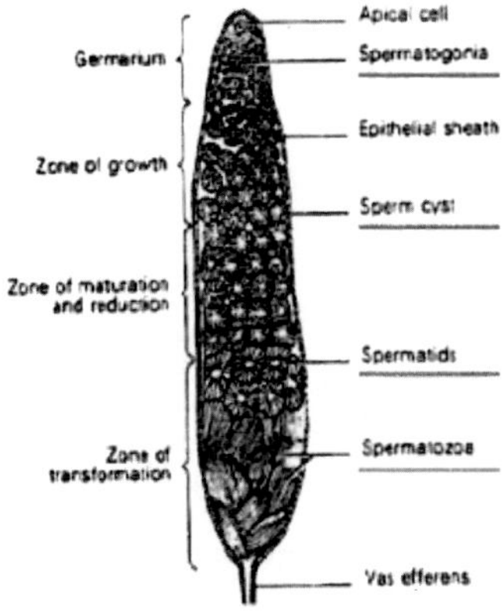

Apical cell
Germarium
Spermatogonia
Epithelial sheath
Zone of growth
Sperm cyst
Zone of maturation and reduction
Spermatids
Zone of transformation
Spermatozoa
Vas efferens

Terminal filament
Germarium
Vitellarium
Follicle
Pedicel

Panoistic
Telotrophic
Polytrophic
Germarium
Oocyte
Vitellarium
Terminal filament
Trophic tissue
Germarium
Nutritive cord
Oocyte
Vitellarium
Follicular epithelium
Terminal filament
Germarium
Follicular epithelium
Oocyte
Vitellarium
Trophocytes or nurse cells
Oocyte with yolk spheres

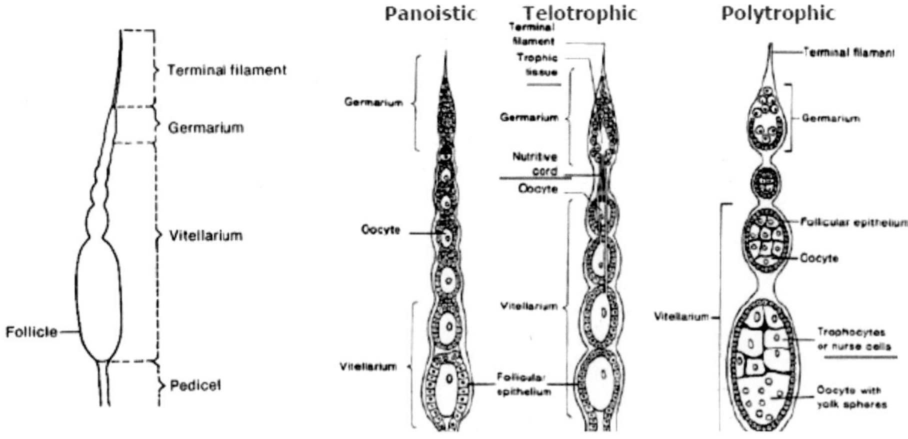

Internal Fertilization

- Males deposit their sperm inside a female's body during an act of **copulation**. This novel adaptation ensured that more sperm found their way to a receptive female
- Pterygota generally:
  - Direct transfer
  - Requires copulatory structures
- **Exception: traumatic insemination**
  - Cimicidae (bed bugs)
  - Penis inserted through body wall
  - Sperm delivered into hemolymph

Male Reproductive System

- **Functions**
  - Sperm production
  - Sperm transfer

The male's reproductive system contains a pair of testes, usually located near the back of the abdomen. Each testis is subdivided into functional units (called **follicles**) where sperm are actually produced. A typical testis may contain hundreds of follicles, generally aligned parallel to one another. Near the distal end of each follicle, there are a group of germ cells (**spermatogonia**) that divide by mitosis and increase in size to form **spermatocytes**. These spermatocytes migrate toward the basal end of the follicle, pushed along by continued cell division of the spermatogonia. Each spermatocyte undergoes meiosis: this yields four haploid **spermatids** which develop into mature **spermatozoa** as they progress further along through the follicle.

Mature sperm pass out of the testes through short ducts (**vasa efferentia**) and collect in storage chambers (**seminal vesicles**) that are usually little more than enlarged sections of the vasa. Similar ducts (**vasa deferentia**) lead away from the seminal vesicles, join one another near the midline of the body, and form a **single ejaculatory duct** that leads out of the body through the male's copulatory organ (called an **aedeagus or penis**).

The male holds the female with the help of a pair of 'claspers'. One or more pairs of **accessory glands** are usually associated with the male's reproductive system. These are secretory organs that connect to the reproductive system by means of short ducts -- some may attach near the testes or seminal vesicles, others may be associated with the ejaculatory duct. The glands have two major functions:

1. Secretion of **seminal fluid**, a liquid medium that sustains and nourishes mature sperm while they are in the male's genital system.
2. Production of **spermatophores**, pouch-like structures (mostly protein) that encase the sperm and protect them as they are delivered to the female's body during copulation.

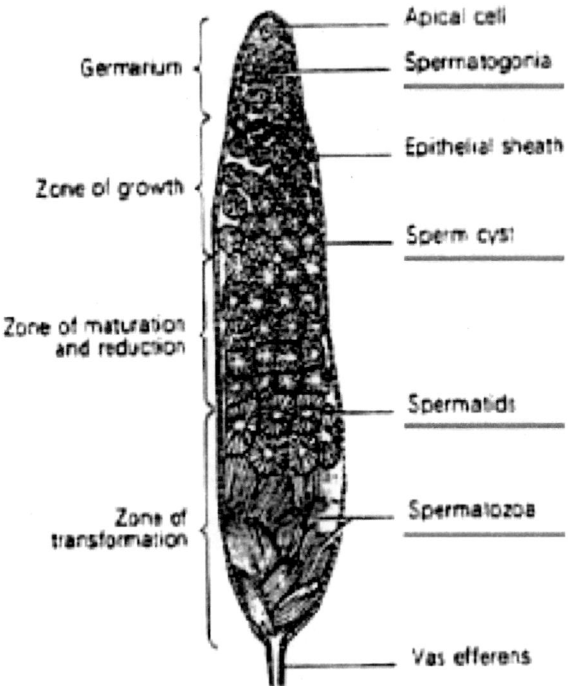

Female Reproductive System

- Functions
  - Egg production
  - Egg fertilization
  - Includes sperm storage
  - Egg placement
  - Oviposition

The female's reproductive system contains a pair of **ovaries**. Each ovary is subdivided into functional units (called **ovarioles**) where the eggs are actually produced. A typical ovary may contain dozens of ovarioles, generally aligned parallel to one another. Near the distal end of each ovariole, there are a group of germ cells (**oogonia**) that divide by mitosis and increase in size to form **oocytes**. During active oogenesis, new oocytes are produced on a regular schedule within each ovariole. These oocytes migrate toward the basal end of the ovariole, pushed along by continued cell division of the oogonia. Each oocyte undergoes **meiosis: this yields four cells -- one egg and three polar bodies**. The polar bodies may disintegrate or they may accompany the egg as nurse cells.

As developing eggs move down the ovariole, they grow in size by absorbing yolk (supplied by adjacent **trophocytes or nurse cells or accessory cells**). Thus, each ovariole contains a linear series of eggs in progressive stages of maturation, giving the appearance of a "chain of beads" where each bead is larger than the one behind it.

Mature eggs leave the ovaries through short lateral duct. Near the midline of the body, these lateral oviducts join to form a common duct which opens into a genital chamber called the **bursa couplatrix**. Female **accessory glands** (one or more pairs) supply lubricants for the reproductive system and secrete a protein-rich egg shell (chorion) that surrounds the entire egg. These glands are usually connected by small ducts to the common oviduct or the bursa copulatrix.

During copulation, the male deposits his spermatophore in the bursa copulatrix. Peristaltic contractions force the spermatophore into the female's spermatheca, a pouch-like chamber reserved for storage of sperm. A spermathecal gland produces enzymes (for digesting the protein coat of the spermatophore) and nutrients (for sustaining the sperm while they are in storage). Sperm may live in the spermatheca for weeks, months, or even years!

During **ovulation**, each egg passes across the opening to the spermatheca and stimulates release of a few sperm onto the egg's surface. These sperm swim through the **micropyle** (a special opening in the egg shell) and get inside the egg. **Fertilization** occurs as soon as one sperm's nucleus fuses with the egg cell's nucleus. **Oviposition** (egg laying) usually follows closely after fertilization. Once these processes are complete, the egg is ready to begin embryonic development.

Externally the sexual organs, called genitalia, of a female insect generally consist of an 'ovipositor' which is often encased in a pair of filaments called a 'sheath' and is which is used to by the female to to lay eggs. The median part of the oviduct which receives the aedeagus during mating is called the 'vagina'.

Ovary types

Panoistic

- No special nutritive/nurse/trophocytes cells
- Oocyte + follicle cells

- Primitive type
- Apterygota, Paleoptera, most Orthopteroid, Thysanoptera, Siphonaptera

Meroistic

- oocyte + trophocytes (nurse cells)
- **Polytrophic**
  - trophocytes accompany oocyte into vitellarium
  - each oocyte has several trophocytes with in its follicles
  - Lice, Dermaptera, Psocoptera, most Endopterygotes
- **Telotrophic/Acrotrophic:**
  - trophocytes remain in germarium, connection to oocyte via nutritive cord
  - Most Hemipteroid, Coleoptera (Polyphaga)

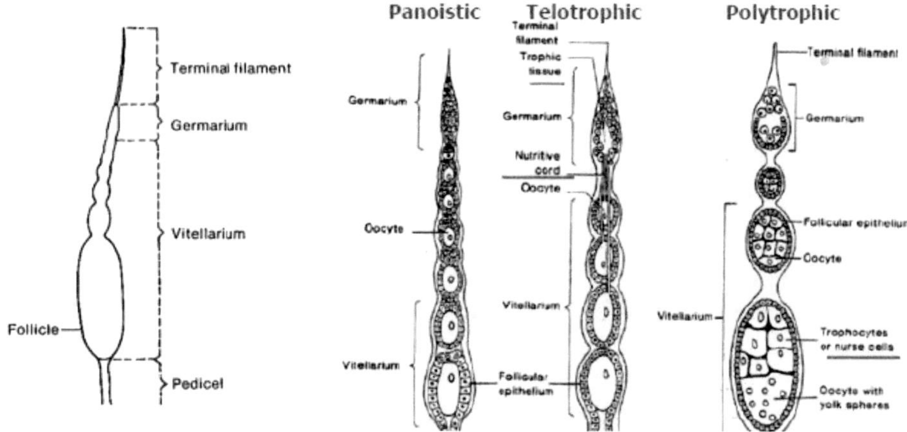

Oogenesis and Ovulation

- Oogenesis occurs in ovariole
- May be continuous process
- May occur once in the life cycle
  - Mature oocyte exits ovariole - ovulation
  - Oocyte passes through oviducts
  - Fertilization in genital chamber

Oviposition

- Gonopore position variable
- Gonopore may be surrounded by ovipositor or other modifications,
  - Often the structure is telescopic
- Oviposition behavior variable
  - Phasmids simply drop eggs
  - Most place eggs 'carefully'
  - Often protected site or on/in food source
  - Some species guard eggs & larvae
  - Earwigs, some hemipterans, social insects

Parental Care

- Wasps (*Eumenes*), Bees (*Megachile*), Dung beetle (*Heliocopris*) : These constructs brood nests, where parent cares for the young to some extent.
- Social insects (honeybees, termites, ants) gave attention to first batch of eggs laid by mother
- Direct maternal protection :
  - *Chrysocoris* bugs, *Gryllotalpa*: guarding the egg cluster
  - Male protecting eggs on the dorsum: *Belostoma*

Inhibition of reproduction in social insects

In honeybees the queen produces secretion from mandibular glands known as Queen substance (9-oxo-trans-2-decenoic acid) inhibits the ovaries of worker bees.

Reproductive Strategies in Insects

Oviparity

- Egg laid shortly after fertilization
- No retention
- No nutrients to embryo after fertilization
- majority of insects have this

Ovoviviparity

- Eggs retained until embryogenesis complete
- Embryo fed by egg reserves
- Female deposits nymph/larva

Viviparity

- Eggs retained and Embryo fed by mother
- Immatures may complete development before deposition The embryo completes development inside the egg while inside the mother and comes out as larva or nymph
- Larviparous=flies (sarcophaga sp) also family Tachnidae)
- Nymphiparous=aphids
- Pupiparous=tsetse flies (immediately burrows and forms a cocoon)

Adneotrophic Viviparity

- Specialised structures are developed from mother to nourish the growing embryos.
- Fully developed egg with chorion moves in the specilaised uterus.
- The emerging larva lives in the uterus and branched maternal glands provide nutritive secretion.
- The larva emerges out of the mother and the non feeding pupal stage is external.
- E.g. *Glossina, Hippobasca*

Haemocoelic viviparity

- Oviducts are absent and mature eggs migrate in the haemocoele
- Embryo lies in the haemocoele
- Eggs may develop without a yolk and shell being nourished by mother
- Eggs may freely float in the haemocoel, in this case the larva may eat their way out, Killing the mother.
- E.g. Gall midges and Strepsiptera

Pseudoplacental viviparity

- The egg is without yolk or chorion.
- Placenta like structure developed either from the embryo or mother or from both.
- E.g. *Hemimermis, Macrosiphum*
- In *Diploptera,* Pleuropodia helped in nutrient absorption.

Paedogenesis

- Reproduction by larval insects. Reproduction by immature or larval stage (asexual, examples among insects involve parthenogenesis)
- e.g.,- several genera of Cecidomyiidae (Diptera) - one species, *Miastor metraloas*, occurs under bark, daughter larvae produced inside female and consume her and escape.
- *Micromalthus debilis* (Coleoptera: Micromalthidae) - only member of family; larvae found in decaying logs, oviparous and viviparous *parthenogenic* larvae.
- In *Midge Henria*, the pupae paedogenetically reproduce to form normal adults and Hemipupa (withstand extreme environmental conditions)

Parthenogenesis

- Development without fertilization
- No need for male
- E.g. Phasmid (*Carauius*)
- Unfertilized eggs produce:
- Males (**arrhenotoky**) in Hymenoptera.
- Females (**thelytoky**) : here females are produced from unfertilised eggs. There are fulcutative and obligatory thelytoky.
    - Falcutative= occurs only under certain conditions
    - Obligatory=means females are obliged to reproduce parthenogenically producing only females. In such organisms males are rare or sometimes even unknown.
- Both **(amphitoky)** in aphids, some wasps, reproduction where eggs undergo a non normally attained haploidy. Males and females are produced.

Cyclic parthenogenesis (Heterogony)

- In this case the insects alternate between amphitoky and obligate thelytoky.
- There is regular alternation of generation during the growing season the females reproduce parthenogenitically to take advantage of the available resources as such an increase in number quickly. In the non growing season (cold & dry) food becomes limiting as such normal reproduction is used.
- Common in aphididae.

Polyembryony

- Found in some endo-parasitic groups only.
- Very common in Hymenoptera (*Platygaster*, several Ichneumonids, Ceccidomyids)
- Advantageous for rapid increase in population
- Single egg results in 2 to 'several thousand' larvae
- Some larvae may be 'defender morphs'
  - Hatch more quickly
  - Eliminate rival parasites
  - Fail to pupate & they die
- Remaining larvae become 'reproductive morphs' that complete development and reproduce to carry on the species

Functional Hermaphroditism

- Resulting from abnormal development
  - Gynandromorphs (genetic mosaics)
  - Intersexes (genetically uniform, result from unstable development)
- Normal development in some scale insects
  - Female scales have both male and female gonads
  - Eggs are self-fertilized
  - In some scale insect species, males are rare

12

Sense Organs

The sense organs in an insect body are distributed on different parts and respond to a given stimulus such as light, sound, touch, chemicals etc.

The sense organs may be classified as

- Visual organs (or) photoreceptors
- Auditory organs (or) organs of hearing
- Chemoreceptors which respond to chemicals
- Tacticle receptors which respond to touch
- Gustatory receptors which respond to taste.

Photoreceptors

Light is perceived by insects through a number of different sense organs and these are the organs that react with light and useful for vision. These are of different types, but most important are the compound eyes.

Compound eyes

- Most insect adults have a pair of compound eyes, one on either side of the head, which bulgeout to a greater or lesser extent, so they give a wide field of vision in all directions.
- Compound eyes are absent in Protura & Diplura.
- Compound eyes are absent in larval forms of holometabolous insects, but have simple stemmata on each sides of the head, and the vision is by ocelli.
- The compound eyes are strongly reduced or absent in parasitic groups such as Siphunculata and Siphonoptera, and in female coccids.
- Compound eyes occur on the side of the head and are quite separate from each other in majority of insects (dioptic eyes), but in some insects, the compound eyes are very close to each other (holoptic eyes). An insect with completely separated compound eyes is the bee (Hymenoptera), while the insect with very closer compound eyes is dipteran flies.
- Each compound eye is an aggregation of similar units known as ommatidia,

the number of which varies from one (in worker ant of Ponera) to >10,000 (dragon flies).

- When only few ommatidia are present, the facets which they present to outside be separated from each other by narrow areas of cuticle; more usually, with larger numbers, the facets are packed close together and assume a hexagonal form.
- If the number of ommatidia is more, these are closely packed and are Hexagonal in shape.
- If number is less, these are loosely packed and are Circular in shape.
- Ommatidia vary in size from insect to insect.
- These develop embryonically in exopterygotes however post-embryonically in endopterygotes.

Structure of ommatidium

- Each ommatidium consists essentially of an (1) optical, light gathering part and a (2) sensory part, perceiving the radiation and transmitting it into electrical energy.
- The optical part of the system usually consists of two elements, a cuticular lens (corneal lens) and a crystalline cone.

The cornea (The Lens)

- This is the outermost transparent colorless layer of cuticle forming the external facet and acting as a lens.
- It is biconvex.
- Like the rest of the cuticle, it is secreted by epidermal cells, each lens being produced by two cells, the corneagen cells, which later become withdrawn to the sides of the ommatidium and form the primary pigment cells.

Crystalline cone

- Beneath cornea are four cells, the Semper cells, which, in most insects, produce the crystalline cone (which is comprised of four cells called ‘Semper cells’ after the man who first described them).
- Normally this functions as a secondary lens.
- This is hard, clear intracellular structure bordered laterally by the primary pigment cells.

- Eyes in which the crystalline cone is present are called eucone eyes. (In some eg. Elatridae and Lampyridae, the Semper cells do not form the crystalline cone, but they extend to the retinula cells as slender refracile strands, and this is known as exocone eyes condition.)

Primary pigment cells (Primary skin cells)

- These are densely pigmented, commonly two in number present around the crystalline cone.
- These are primarily corneagen cells and after formation of cornea, they become primary pigment cells withdrawing to the sides of ommatidia, surrounding crystalline cone.

Retinular cells (nerve cells)

- Immediately behind the crystalline cone in eucone eyes are the sensory elements.
- They are elongate nerve cells known as retinula cells.
- The receptive parts of an insect's eye are the 'retinula cells'.
- The cytoplasm of the retinula cells contains pigment granules, which are especially concentrated at the edge of the rhabdomere, but those granules do not contain the visual pigment.
- Each ommatidium normally has eight retinula cells arranged to leave a central core space in the centre of the ommatidium, into which each retinula cell projects a series of microvilli (like very small fingers).
- The microvilli from each retinula cells are the actual light detecting part of the cells and are collectively referred to as the rhabdomere (think cornea).
- The rhabdomeres are the microvilli emanating from retinula cells into central gap between all eight retinular cells
- The eight (or occasionally 7 or 9) rhabdomeres (sets of microvilli) form a rhabdom.
- The retinula cells are connected to axons at the base of the eye, it is these which carry the information collected by the lenses and converted into electrical impulses by the rhabdom to the brain, thus allowing the insect to see.
- The corneal lens is supported by 'primary pigment cells'.
- The retinula cells and associated rhabdoms are supported by 'secondary pigment cells'.

Variation in the lens system

- Eyes in which the crystalline cone (produced by 4 semper cells beneath cornea) is present are called eucone eyes.
- In some eg. Elatridae and Lampyridae, the simper cells do not form the crystalline cone, but they extend to the retinula cells as slender refracile strands, and this is known as exocone condition.

Variation in the retinula cells

- The arrangement of the retinula cells falls broadly in two classes. (1) those in which the rhabdom extends to the crystalline cone, and (2) those in which there is a clear zone between the cone and rhabdom.
- Eyes in which there is clear zone between cone and rhabdome are called as clear-zone eyes; they were previously known as superposition eyes, on the basis of their functioning. These eyes are found in crepuscular and nocturnal insects.
- Eyes in which rhabdom reaches the crystalline cone are called as apposition eyes.These are found in diurnal insects.
- Superposition eyes can achieve greater sensitivity than apposition eyes, so are better suited to dark-dwelling creatures.

Simple Eyes (Ocelli)

- Simple eyes or Ocelli are present in most insects to some degree.
- Two different forms of ocelli have been described for insects, dorsal ocelli and lateral ocelli.
- Dorsal ocelli are found in adult insects and the larvae of hemimetabolous insects. Typically there are three, forming an inverted triangle antero-dorsally on the head. Dorsal ocelli occur mostly in adult insects and are situated on the front of the insects face in the area of the 'frons' and or the 'epicranium'.
- Lateral ocelli generally occur on the sides of the insect head and are the form of eye most common in larval forms.
- Ocelli very sensitive to low light levels.
- Ocelli have role in circadian rhythm (day light responses eg. In diapause)

Dorsal ocelli

- Generally present in all adult insects and nymphs of hemi-metabola.
- Usually number is three though some times reduced to two, arranged in triangle on vertex and frons.
- These are highly reduced and represent by fenestrae in cockroaches.
- Ocelli absent in Apterygota forms and present in winged insects.
- Typical ocelli have a single thickened cuticlular lens, but in other cases the cuticle is transparent, but not thickened. The space beneath the cuticle being occupied but with transparent cells. Though the lens is optically capable of forming an image, it does so at a level far below the retina. Blackening of the dorsal ocelli reduce the speed with which some insects respond to stimulation of the compound eye by light. They are therefore regarded as stimulatory organs.
- Dorsal ocelli develop embryonically or post-embryonically.
- All parts derived from Ectoderm (ectodermal in origin).

Lateral ocelli or stemmata

- Lateral ocelli are only the eyes present in insect larvae of holometabola and in apterygote adults.
- Ocelli have very poor resolution power.
- Caterpillars behaviorally scan from side to side using these visual organs. Similar to structure and function to ocelli. These are present on either side of the head in the position corresponding with those of the compound eyes. The number may vary from 1-7 or more on each side. Probably not image forming but sensitive to direction of light.
- These are innervated from the optic lobes of the brain.
- Stemmata can form an inverted image irrespective of the distance of the object.
- Each stemmata can form image of only one visual field of object thus forming a very coarse mosaic of intensities. By moving the head from side to side it can examine a arger field and is at least capable of orientating it self towards the boundary between light and dark parts of the environment.
- Caterpillar behavior for food and pupation suggest that they have color vision.

- Start development during embryonic stage but the process may only be completed during post embryonic life.
- All parts derived from Ectoderm (ectodermal in origin).

Types of Image formation

- Iris cell pigments absorb light that is not refracted. Their structure differentiates the superposition eyes from appositions eyes.
- **Aposition** eyes have ommatidia optically isolated, rhabdom extends/ reaches to crystalline cone and function best under bright light conditions. These eyes are found in diurnal insects.
- In **superposition** eyes, the secondary iris cells have mobile pigments which can isolate or expose retinular layer of ommatidium to light from adjacent ommatidia (optically not-isolated). In these eyes, there is a clear zone between the cone and rhabdom (clear-zone eyes). These are found in crepuscular and nocturnal insects. These eyes are adopted for dim light conditions.
- Eyes adopted for bright light conditions are apposition eyes, and eyes adopted for dim light conditions are superposition eyes

Auditory Receptors

- Insects are provided with structures (or) organs that are able to perceive the sound waves (or) the aquatic water currents. Among the organs of hearing, the auditory hairs, tympanal organ and Jhonston's organ are important.
- Delicate tactile hairs: Present in plumose antenna of male mosquito.
- Tympanum: This is a membrane stretched across tympanic cavity responds to sounds produced at some distance, transmitted by airborne vibration. Tympanal membranes are linked to chordotonal organs that enhance sound reception. Tympanal organs are located
  - Between the metathoracic legs of mantids.
  - The metathorax of many nectuid moths.
  - The prothoracic legs of many orthopterans.
  - The abdomen of short horned grasshopper, cicada.
- The wings of certain moths and lacewings. Sounds are produced by many insects (Orthoptera) using variey of mechanisms. Some sounds, such as that resulting from vibration of the wings in flight. Many but not all insects can hear sounds; some even hear sounds that we can't hear ourselves. Insects

hear through one of four different ways, the most common of which is the tympanum.

Tympanal organs

- Always occur as paired organs.
- They are composed of a thin cuticular membrane (the tympanum) stretched across an air space of some sort and some form of connection to the nervous system. Scoplopodium form the tympanal organ of the locust.
- In the Orthoptera (Grasshoppers and Crickets) tympanum are common, though situated in different places in different species, i.e. on the **first abdominal segment in Grasshoppers** and on the **front legs (base of fore tibia) in the Crickets**.
- Tympanal organs also occur in the Cicada (Cicadidae, Hemiptera) and some families of the Lepidoptera, (i.e. Noctuidae, Geometridae, and Pyralididae).
- In Lepidoptera these are present on thorax or abdomen

Johnston's Organ

- The second segment of antenna, pedicle, usually contains a special sensory structure auditory organ, known as Johnston's organ, which is absent in Diplura, Collembola.
- It detects the motion in the flagellum (third segment)
- The Johnston's organ in male mosquito is extremely important in mating. The male detects the wing beat frequency of female and is attracted to her.
- This can also sense wind.

Auditory Hairs

- These are present on the body of insects such as larvae of Lepidoptera as well as on some Orthoptera which are developed from the modified epidermal cells. These respond to the sounds of air (or) water currents mediated by the **hair sensillae** (or) **trichoid**

The Pilifer

- This is a unique auditory organ found only in the head of certain species of Hawk Moths of the subfamily Choerocampinae its optimum frequency is between 30 and 70 kHz which would allow it to hear the echolocation calls of many of the larger insectivorous bats.

Other sense organs for detection various stimuli

Hygroreceptors

- Sensilla that response to changes in the humidity and are extremely important to insects that need to conserve water loss. Sense cells responsive to temperature and humidity are present in the same sensillum. All insects have this on antenna.

Osmoreceptors

- The ability to respond the changes in osmotic pressure of the fluid. Osmoreceptors have not been studied in insects in detail. Literature suggests the presence of these receptors in horsefly.

Thermoreceptors

- Heat reception is used by hematophagous insects (blood sucking) to induce biting.
- These are present in poikilothermic insects and sensitive to temperature changes. In bed bug it is useful to locate the host utilizing the temperature gradient of the host. Two hairs on the tarsi of foreleg of Glossina mosquitoes, on tarsi of *Periplanata Americana*, found in heat seeking Melanophila beetles.

Chemoreceptors

- Detect chemical energy. Insect chemoreceptors are sensilla with one pore (uniporous) or more pores (multiporous).
- Uniporous chemorceptors mostly detect chemicals of solid and liquid form by contact and are called as **gustatory receptor**. Many sensor neurons located in antenna are of this type.
- Multiporous chemoreceptors detect chemicals in vapour form, at distant by smell and are called as **olfactory receptor**. Few sensory neurons located in trophi and tarsi are of this type. Each pore forms a chamber known as **pore kettle** with more number of pore tubules that run inwards to meet multibranched dendrites.
- Chemoreceptor is sensitive to chemicals, stimulation by chemicals can occur in the following different ways:

Olfactory or smell chemoreceptor

- They provide sense of smell. Mechanism of perception of chemicals in gaseous state at high concentration is known as olfaction. The antennae are the primary site of olfactory organs and often bear many thousands of these structures. The mouthparts also carry olfactory structures in many species.
- The sense of smell or response to volatile chemicals, i.e detection of chemicals by smell, can be located anywhere antenna, palps and genetalia. Pheromone binding protein is produced by the trichogen and tormogen cells on antenna.

Gustatory or contact chemoreceptor

- They provide sense of taste. Mechanism of perception of chemicals in liquid state at high concentration as known as contact chemoreceptor.
- **Gustation or contact reception** : Detection of chemicals by contact, can be located anywhere on the body, but usually found on labrum, maxillae, labium, antenna, tarsi, ovipositor. The antennae (Hymenoptera), Tarsi (many Lepidoptera, Diptera, and the honeybee), Ovipositor (parasitic Hymenoptera and some Diptera) and General body surface.

Form of Chemoreceptor

Chemoreceptor consists of one or more neurons associated with cuticular structures having small pores through which molecules enter and stimulate dendrites. Cuticular structures have wide variety of forms.

- Trichoid sensilla: Sensory cells provided with hairs.
- Sensilla basiconica: Sensory cells provided with peg-like or cone-like structure.
- Sensilla coeloconica: Sensory cells provided with pits.
- Sensilla placodea: Sensory cells provided with plates.

Functions of chemoreceptor

- Chemoreception, essentially taste (gustation) and smell (olfaction), is an extremely significant process in the selection, as it initiates some of their most important behaviour patterns that is section of food, oviposition site, location of host or mate, and responses to commercial attractants and repellents.

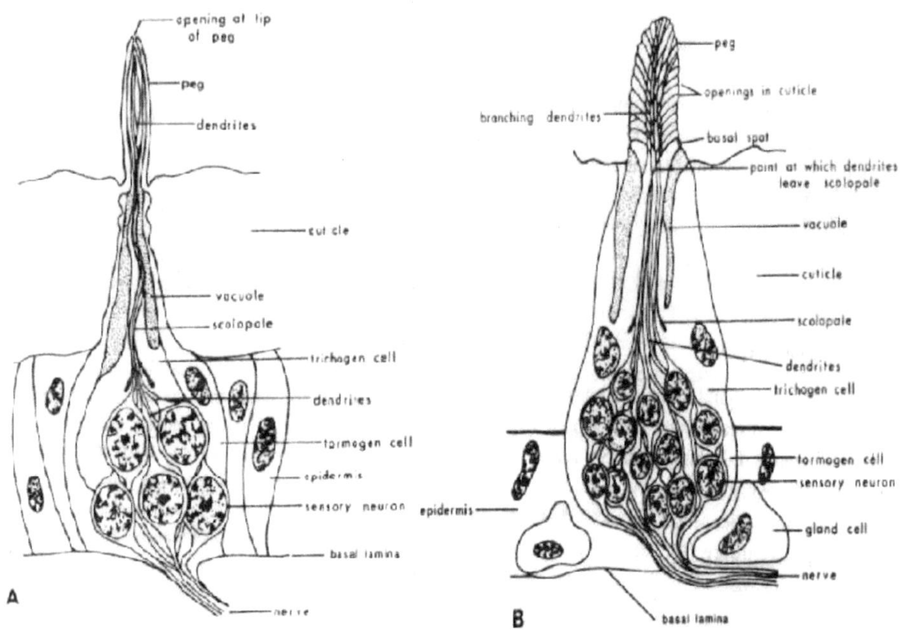

Mechano receptors

Trichoid sensilla

- Hair like little sense organ. Sense cell associated with spur and seta. These cells are sensitive to touch and are located in antenna and trophi (mouth parts).
- The trichoid sensilla are the simple articulated sensory hairs and distributed on the entire body surface and commonly called as the sensilla.
- They are made up of cuticle and articulated within the socket with the body wall. The trichoid sensillum is formed by two cells; the hair by the trichogen cell and the socket by the tormogen cell. Both the cells are modified epidermal cells. Each sensillum is innervated by a bipolar sensory neuron. The dendrite of the neuron is enclosed at the base of the hair by a cuticular tubular sheath, called the scolopale. It is, in some case, provided with a distal cap, called the scolopale cap.
- Trichoid sensilla present on the antennae, tarsi, tibia and cerci. These organs of the insects sub-serving the sense of touch.

• **Functions**

- There are some trichoid hair plates at the joints of various appendages and function as proprioceptors during sliding of the segments over each other.
- There are tactile hair- beds on the facial region of the head of locusts and Lepidoptera, on the neck of dragonflies, on the wing margins of Lepidoptera which are responding to the air movements during flight.
- The tactile hairs of the antennae and lower segments of legs perceive earth-born vibrations in terrestrial insects and water surface vibrations in aquatic insects.

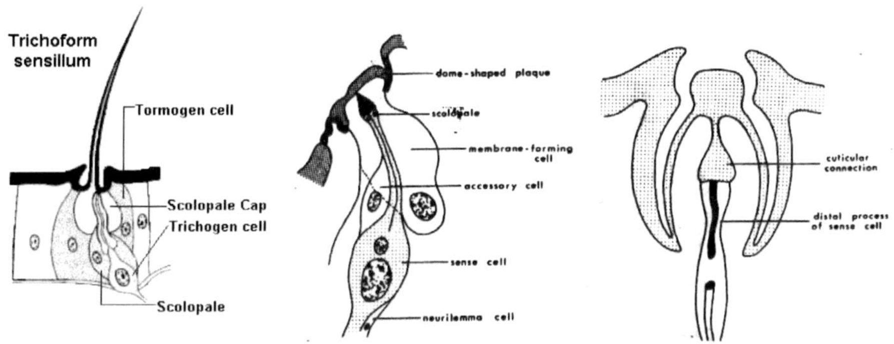

Trichoid sensilla **Campaniform sensilla**

Campaniform sensilla (Dome sensilla)

- Terminal end of these sensilla is rod like and inserted into dome shaped cuticula. These cells are sensitive to pressure and located in leg joints and wing bases.
- These can not be seen externally but recognized from dome shaped cuticular areas. They elevated above or depressed below the general body surface. Cell structure and arrangement is similar to that of trichoid sensilla.
- They occur in various parts of the body, wing-base, halteres, cerci, palps and on the base of trochanter, femora, tibia and tarsal segments.
- The cuticular stretches and the forces exerted on the cuticle by muscles and gravitational forces stimulate the campaniform sensilla. The campaniform sensilla function as the proprioceptors. They respond to the mechanical stimuli, in terrestrial insects, water pressure in aquatic insects and air pressure in flying insects.

Chordotonal organ

- The specialized sensory organs that receive vibrations are subcuticular mechano receptors called chordotonal organ. An organ consists of one to many scolopidia, each Chordotonal organ consists of single unit or group of similar unit is called scolopidia. They are sub-cuticular and are attached to the cuticle at one or both end often no sign of their presence. Each scolopidia consists of three cells:

  a. Neuron

  b. Scolopale cell or enveloping cell

  c. Cap cell

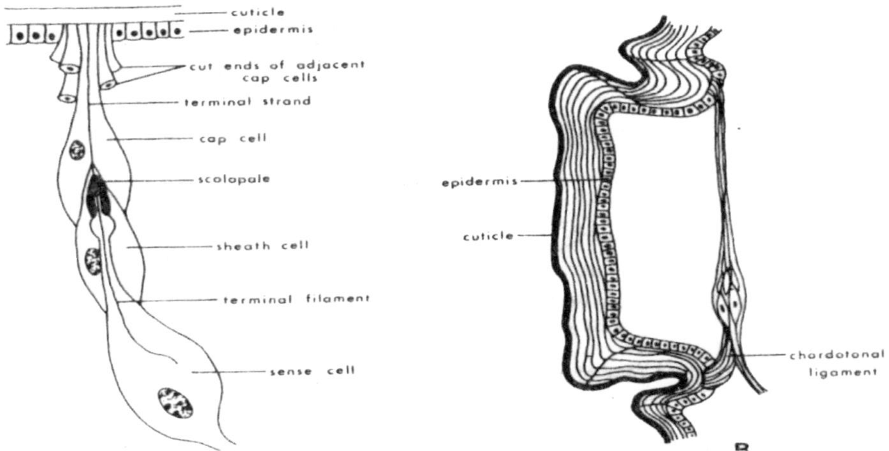

Chordotonql organs

- The chordotonal organs occur in legs at femoral, distal tibial and tibio-tarsal regions, in abdomen and wing base.
- They are stimulated by passive movement of segments, tension of muscles, internal pressure changed due to blood and tracheae.
- Chordotonal organ located in the proximal tibia of each leg, used to detect substrate vibration. Subgenual organs are found in most insects, except the Coleoptera and Diptera.

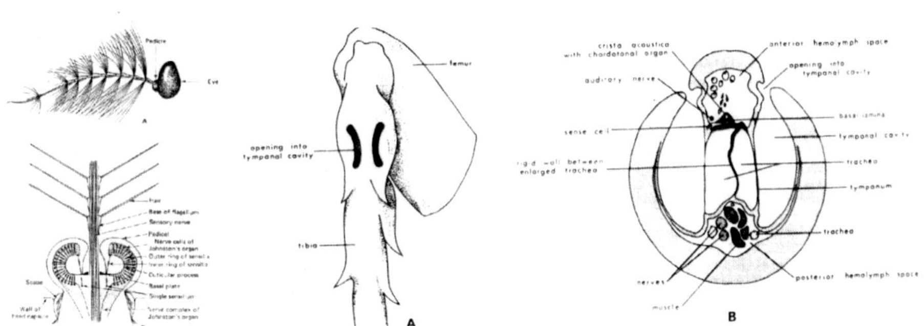

Johnston organ **Tibial tympanal organ**

Specialized chordotonal organ

a. Johnstons organ

b. Auditory or tympanal organ

a. Johnstons organ

- Jonstons organ is a specialized chordotonal organ in the 2nd antennal segment, occurs in all adult insect except Collembola and Diplura. It consists of single mass or several groups of scolopidia and is highly developed in Culicidae, where the pedicel is enlarged. Axon of sense cells run back and enter the antennal nerve. It perceives movement of antennal flagellum and flight speed indicator.
- All adults insects and many larvae have a complex chordotonal organ called Johnston's organ lying within the second antennal segment (Pedicel). These organs sense movements of antennal flagellum. It also functions in hearing in some insects like male mosquitoes and midges.

b. Auditory or tympanal organ

- Tympanal organs are present in the adult of many insect species. Tympanal organ consists of a thin layer of cuticular structure, called tympanic membrane, air sac and a group of chordotonal organ. Tympanic membrane and air sac form drum, sound waves that strike the drum cause it to vibrate and therefore the sensilla to be stimulated.

Proprioception (positioning of their body parts in relation to the gravity).

- In desert locust, they have specialized hairs with swollen bulbs at the tip of cerci. Centre of gravity is presumed to be inside the bulb. So, hair sensilla are very important for various senses, spread all over the body in various forms for smell, touch, gravity, pressure, taste, pheromones etc. Sense of touch sensilla found on allover the body, antenna, cerci, ovipositor, mouth parts etc., sense of air movement sensilla found on antenna, cerci. Sensitive to sound waves, vibration of substratum and pressure changes.

13

Insect Systematics

History

- ***Aristotle (384-322 BC):*** A good beginning of taxonomy was made by Aristotle (*Historia Animalium*). He erected *entomon*. He could distinguish mandibulates and haustellates; winged and wingless forms. The insect orders like coleopteran, dipteral, psyche (Lepidoptera) were created by him. Hence, he was called as *Father of Biological Classification.*
- ***Linnaeus (1707-1778):*** *Father of Taxonomy and Binomial Nomenclature. He coined the term Insecta.* He used the term *Systematics*. His greatest contribution is *Systema Naturae.*
- Brawer (1885): Laid foundations for modern system of classification and responsible to divide class Insecta into Apterygota and Pterygota.
- Michael Anderson (1727-1806): Contemporary and rival of Linnaeus. Introduced *Numerical Taxonomy* and hence is called *Father of Numerical Taxonomy.*
- Johan Christian Fabricius (1745–1808), the first systematist considered to specialize on insects, published in 1778 his Philosophia Entomologica (40), the first textbook of entomology.
- Fabricius had already set forth his system of insects in 1775 (Systema Entomologiae), his system of genera in 1776 (Genera Insectorum), and then numerous works on species, culminating in individual accounts of the species of each order

Taxonomy (Greek: Taxis: Arrangement and nomos: law).

- Term first given by De Condolle (1813) and is defined as the theoretical study of classification, including its basis, principles, procedure and rules (Simpson,1961).
- Taxonomy is the science of classification. It can be defined as placing biological organisms or forms in order.
- The word finds its roots in the Greek word, *taxis* (meaning 'order', '*arrangement*') and *nomos* ('law' or 'science'). Taxonomy uses taxonomic

units, known as *taxa*

- (singular taxon). It was coined by *A P De Candolle*, a professor of Montpellier University in France.
- Taxonomy is the theory and practice of classifying organisms into categories based on similarities or differences, and describing of these systematically. Taxonomy is the science of naming organisms.
- Taxonomy is the practice and science of classification.
- Taxonomy is the study of the principles of scientific classification.

Levels of taxonomy

There are three levels of taxonomy.

Alpha taxonomy

It is the level of the taxonomy concerned with the characterization and naming of species.

Beta taxonomy

It is the level of taxonomy concerned with the arrangement of species in natural system in lower and higher taxa.

Gamma taxonomy

It is the level of taxonomy dealing with various biological aspects of taxa ranging from the study of speciation and of evolutionary rates and trends.

Taxonomic categories

The species is the basic unit of taxonomic classification. The hierarchy of generally accepted taxonomic categories are as follows:

Kingdom
Phylum
Subphylum
Superclass
Class
Subclass
Cohort
Superorder
Order
Suborder
Superfamily (ends in 'oidea')
Family (-idae)
Subfamily (-inae)
Tribe (-ini)
Genus
Subgenus
Species
Subspecies

Systematics (Latin: Systemia: Systems of classification)

- Systematics is the area of biology devoted to the *classification of organisms*.
- Originally introduced by Carl Linnaeus (1707-1778), who based his classification system on physical traits, systematics now includes the similarities of DNA, RNA, and proteins across species as criteria for classification.
- *Systematics* is the science of study of kinds of organisms, and of any and all relationships among them, or *science of diversity of organisms.*
- *Systematics* is the study of the classification of organisms, and their historical relationships, also known as phylogenetics.

Classification

- Ordering of organisms into groups or sets on the basis of their relationships. It helps to create a reliable, easy to use filing system for identification on the basis of physiology, ecology, behavior and phylogeny of the organism.
- It is the scheme of arranging organisms into different groups of systematic study. Classification is the arrangement of organisms in to groups on the basis of their relationships.
- Taxonomy is properly the describing, identifying, classifying, and naming of organisms, while "classification" is focused on placing organisms within groups that show their relationships to other organisms.

Purposes of Classification

- Identification of species
- Provide system of information retrieval (Communication)
- To trace phylogenetic relationships.
- Serve as a summarizing and predicting device.

Characters used in taxonomy

1. Homologous characters

   Wings, legs (exception raptorial forelegs in mantid and mantispid)

2. Ancestral characters

Kinds of classification

Cladistic approach

- Certain features are given more weight than others in deciding relationships between group of organism. Certain characters have undergone greater evolutionary modification than others. Family tree is called cladograms.

Phenetic approach

- Establishment of relationship between organisms on the basis of as many features as possible giving all features equal weight. Family tree is called phenogram.

Hierarchy of biological classification

- A *hierarchy* is an arrangement of items (objects, names, values, categories, etc.) in which the items are represented as being "above," "below," or "at the same level as" one another and with only one "neighbor" above and below each level.
- *The biological unit in the process of classification is the individual organism.*
- Closely related similar individuals are grouped into species, closely related species into genus, genera into family, families into order, orders into class, classes into phylum and phyla into kingdom.
- Sometimes there are certain intermediary categories like sub-class, sub-order, superfamily, sub-family, tribe etc.
- The sequence of classification categories with hierarchy can be remembered by memorizing this mnemonic phrase: King Phillip, Comes Out For Goodness Sake!

Identification

All insects present in the world are not yet identified and described (about 25% of the insect species are unidentified). An already described species produce new race which requires further identification. Identification helps to understand about the organism and to take proper control measure. Insects can be identified through an expert, by comparing the available collections, using photographs and pictures and by using taxonomic keys. Key is a tabular statement presenting alternatives, describing about the features of an organism. Most of the keys are **dichotomous** i.e., always dividing into two or they are always in the form of couplets and give a clear cut alternative. Keys can be constructed based on single character **(monothetic key)** or many characters **(polythetic key).** Polythetic key is more advantageous. Monothetic key has three disadvantages (i) the organism may be an exception for a particular character (ii) chances of erring is more (iii) if the particular body part is broken on which the key is made, then the key cannot be used. Key can be classified based on evolutionary principles also as **phylogenetic key** and **arbitrary key**.

(i) Phylogenetic key: The key is based on the evolutionary relationship. The group appears only once in the phylogenetic key.

(ii) Arbitrary key: The taxa or group appears at several places in the key. It has more advantages.

Nomenclature

(Latin Nomen- name and Calare- to call) literally means call be name. Nomenclature is involved with naming of organisms and groups of organism, rule and procedures to be followed in such naming. Since 1901, International Code of Zoological Nomenclature is followed. Current rules are revised in 1964. Binomial nomenclature was introduced by Carolus Linnaeus. The first edition of the Carol Linnaeus's ***Systema Naturae*** was published in 1735. In this Linnaeus published the System of nomenclature. Binomial Nomenclature was published in the 10th edition (1758).

Common name

- They are **inaccurate** because it varies from region to region and country to country and there is **no uniformity** followed in naming the organisms.

  e.g. Locust is a bug referring cicada in European countries and normally locusts also refer to short horned grasshoppers living in groups.

- **Common name is not available for all organisms.**

  Eg. Squash bugs present in cucurbitaceous plants are represented by many species but no common name is available for each species. It is available only for a large group like order and family.

- **Same common name is used for insects of different orders.**

  eg. Flies. A true fly has only two wings whereas other insects like mayfly, dragonfly etc. are also mentioned as flies.

- **Homonym**- Same name is used for describing two different type of insects.

  e.g. Boll worm is a common term used for more than five species of boll feeding insects.

- **Synonym**- More than one name denoting a single insect.

  e.g. Gram pod borer, American bollworm denotes *Helicoverpa armigera*.

Scientific Nomenclature

The system of naming organisms using two words is called **Binomial nomenclature** (**Trinomial nomenclature** if three words are used). This system of naming gives accurate information. It is universal and accepted in all parts of the world. The rule regarding the naming of organisms is contained in **International Code of Zoological Nomenclature.**

Normally there are two names, the first is the generic name and the second name is the species name. The names e.g. Head louse: *Pediculus humanus capitis.;* Body louse: *Pediculus humanus corporis.*

The first letter of the generic name is in capital and the first letter of the species and subspecies are in small letter. All the words are Latinized and written in italics or it should be written and underlined separately. The authority name is written after the species name. It starts with capital letter. The author name is put in bracket if the taxa has been reclassified and placed in another group. Eg. Moringa fruit fly, *Gitona distigma* (Meigon).

- Scientific nomenclature was started with Linnaeus. He used *binomial system of nomenclature* for the first time for plants and animals in his 10th edition of for the first time for plants and animals in his 10th edition of *systema nature* (1758).
- *Carl Linnaeus*, also known as Carolus Linnaeus (1707-1778), is often called the *Father of Taxonomy*. His system for naming, ranking, and classifying organisms is still in wide use today (with many changes). For Linnaeus, ***species*** of organisms were real entities, which could be grouped into higher categories called ***genera*** (singular, **genus**).

Binomial Nomenclature

- It is the double name of an organism for genus and species in latin, followed by the author name in full without punctuation between them.
- If the sub-species name is also given, it is *trinomial*. The scientific name is printed in *italics (eg. Apis dorsata).* If it is written, it should be underlined (eg. *Apis dorsata*) separately. The genus name is always starts with capital letter and species name with small letter.
- The value of the binomial nomenclature system derives primarily from its economy, its widespread use, and the stability of names it generally favors.
- The same name can be used all over the world, in all languages, avoiding difficulties of translation.

International Code of Zoological Nomenclature (ICZN)

- The International Commission on Zoological Nomenclature is dedicated to achieving stability and sense in the scientific naming of animals, and also publishes *International Code of Zoological Nomenclature* and the *Bulletin of Zoological Nomenclature* containing applications to, and rulings by, the Commission. ICZN will distribute this information as widely as possible, working towards the provision of a free service.
- The ICZN Secretariat is housed in the Department of Palaeontology at the Natural History Museum, London, SW7 5BD U.K.

Law of Priority

- *The underlying principle in ICZN is the law of priority*. It may so happen that *different names* are given by different scientists to the same organism. In such cases the *first published will be accepted as the correct name* provided the author has followed the rules and all the other names go as ***synonyms***. Conversely, the same name may be given by different scientists to different organisms. In such cases, the first published will be accepted as the correct name and the remaining go as ***homonyms*** and they have to be renamed.
- You should note that most animals and plants are known only by their **genus** and **species** names, i.e. *Heliothis armigera*, note also that while the **genus name is spelt with a capital letter the species name is normally spelt with a small letter, and that in printed text the pair are normally written in *italics*** the rest usually remains unsaid.
- You should also note that though an animal will generally share its genus, family, order and class names etc. with other animals its combination genus/ species name will be unique to it.
- In some books you may find a persons name or part of a persons name written after the animals name i.e.*Euclida glyphica* (Linnaeus, 1758) this mean that Linnaeus was the first person to name this moth and that he did so in a work published in 1758, the brackets around his name mean that the species has since been moved to another genus. Linnaeus, being the first person to use this system widely, put a lot of moths in one genus, such as *Noctua*, and later taxonomists realized that some of them belonged to different genera and moved them, like putting them in a different box. This means that the species was named and described by Linnaeus, but the genus was described later be a different taxonomist. This process is called taxonomic revision.
- Also you will occasionally see some species written in a shortened for *Formica exsecta* Nyl., Nyl stands for Nylander, Finally you should

know that scientist are still arguing about some of the family, class and order names so these may be different in different books. Note also that identifying, describing and naming things, i.e. assigning them to particular groups is ***taxonomy***, while arranging those groups in a coherent order which reflects their evolution and relatedness is ***classification***. Another word is ***Systematics*** which may be defined as the study of the diversity of organisms and the way they relate to each other, ***modern Systematics is called Phylogenetic Cladistics*** and has a whole set of special rules telling you how to do it properly.

Analytical Key

- It is a devise used to identify all sorts of organisms. Keys work by a process of elimination, gradually narrowing down the number of possibilities. It is important to understand that a key is much more trustworthy in proving that your insect is not A, than in proving that it is B. It might, for instance, actually be C, a species not mentioned in the key. A key does not prove anything positive, it only suggests possibilities. People used to be faddy about the sort of key they liked, but these days the arrangement that has the greatest acceptability for simplicity and directness is the dichotomous key, so-called because at each step it asks you to choose between two alternatives.

The following is the example:

1a. Insect with wings..2

1b. Insect without wings...27

2a. Insects with four wings (two pairs)..3

2b. Insects with only two wings (one pair)...25

3a. Wings covered with scales............................Butterflies and Moths

Lepidoptera

3b. Wings not covered with scales, though they may be hairy..................................4

4a. Fore-wings partly or entirely horny or leathery and used as covers for hind-wings often much narrower than hindwings...______...5

4b. Both pairs of wings entirley membranous (flexible) and used for flying............12

5a. Mouth-parts tube-like, adapted for piercing and sucking.. True Bugs **Hemiptera**

5b. Mouth-parts adapted for biting and chewing..6

6a. Fore-wings and hind-wings with veins, hind-wings stiffer and harder than and serving as covers for hindwings.................................________......................7

6b. Fore-wings without veins, and modified into hard, horny cases for hindwings…………...10

7a. Body dorsoventrally flattened.................Cockroaches

Dictyoptera; Blattodea

7b. Body rounded or quadrate in section...8

8a. Forelegs raptorial, adapted for grasping and holding........Preying Mantids ...

Dictyoptera;

Mantodea

8b. Forelegs not raptorial...9

9a. Prothorax as large as or larger than meso and meta thorax, hind legs generally enlarged and adapted for jumping........Grasshoppers and Crickets **Orthoptera**

9b. Prothorax smaller than meso and meta thorax, legs normally similar in thickness, if hind legs enlarged then not used for jumping............Stick-Insects

Phasmida

10a. Fore-wings short...11

10b. Fore-wings as long as, or nearly as long as abdomen the 2 wings may be joined where they meet along the animals back and hence never used for flying..Beetles ....

Coleoptera

Species Concept

- It can be defined as the smallest grouping capable of reproduction and the production of fertile offspring." The name of a species is called a scientific binomen. It's a two part Latin name the first of which is the genus and the second is the species, which is always italicized. Examples: *Heliothis armigera* Hubner, *Bemisia argentifolii* Bellows and Perring, *Ceratitis capitatà* (Weidemann).

Species

- It is group of individuals which are similar in structure capable of interbreeding and producing fertile offspring but at the same time reproductively isolated from other such groups.
- Groups of interbreeding natural populations that are reproductively isolated from other such groups.

Subspecies

- It is an aggregate of phenotypically similar populations of a species, inhabiting a geographic subdivision of the range of species and differing taxonomically from other populations of the same species (biotype/race/strain).
- Sub species is a geographically defined aggregates of local populations which differ taxonomically from other such sub-divisions of the sibling species.

Sibling species

- It is a term applied to pair or groups of very similar and closely related species.
- Morphologically similar but reproductively isolated.

Sympatric species

- These are those normally occupying the same geographical area.

Allopatric species

- These are those occupying completely defined area.

Genus

- A group of species having some definite similar characters or relations.
- A taxonomic category containing a single genus species or a monophyletic group of species which are separated from other taxa of the same rank by a decided gap.

Sub-Family

- Is a group of allied genera.

Family

- A taxonomic category containing a single or group of genera of common phylogenic origin which is separated from related families by a decided gap.

Order

- Families showing similar characters from order.

Taxon

- A group of real organisms recognized as a formal unit at any level of hierarchic classification.

ICZN (International Code of Zoological Nomenclature)

- It is the system of rules and recommendations authorized by the International Congress of Zoology. The Code consists of **87 articles.**

Typification of Species

According to ICZN, the author while describing a new species or group has to designate a type.

- ***Type*** is a zoological object on which the original published description of a name is based
- ***Holotype:*** The single specimen selected by the author of a species as its type, or the only specimen known at the time of description; a true type
- ***Allotype:*** It is a specimen of the opposite sex to the type.
- ***Paratype:*** A specimen cited in the original description other than the holotype.
- ***Cotype*** : Any one of all specimens present before the describer at the time the description is drawn (not including type and allotype)
- ***Neotype:*** If holotype is damaged irreparably or lost, the specimen selected by the authority in group to replace is called neoype.\
- ***Homeotype:*** The specimen compared by the authority with holotype and sent out for comparative study.
- ***Genotype***: The species which is designated as the type species of a genus, up on which it is based.
- ***Isotype:*** A duplicate of the type.
- ***Topotype***: A specimen collected in the exact locality from where the original type was obtained.

Steps in arthropod evolution

1. Common ancestor - legless annelid

   - undifferentiated body

   - development of somites, antennal segments, and simple eyes (like Onychophora)

2. Early arthropod - better developed eyes and jointed legs

3. Reduction in appendages
   - cephalization
   -modification of first appendages into mouthparts
   - development of three tagmata

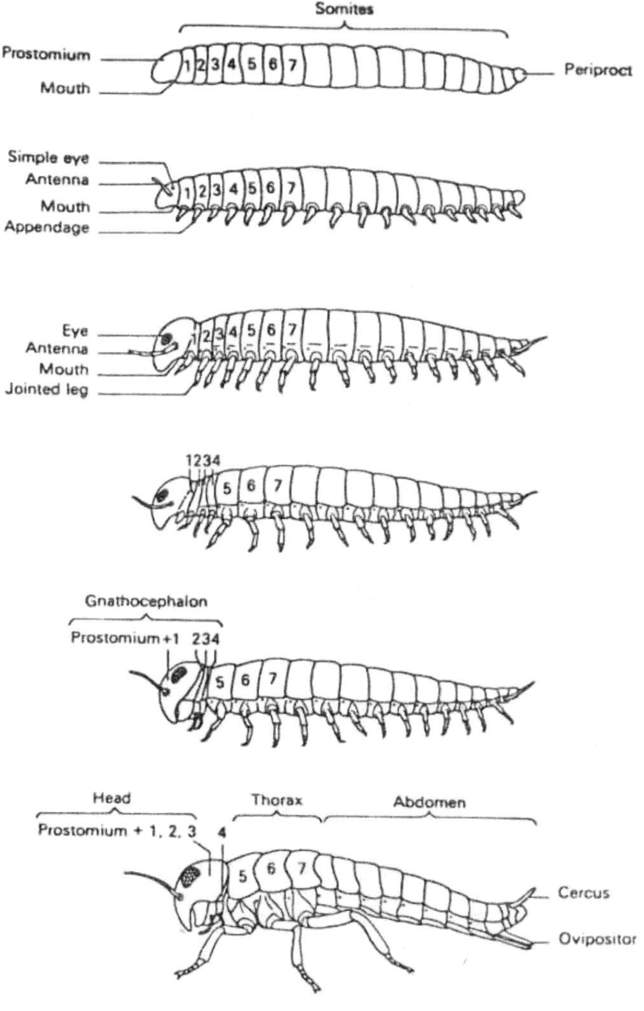

Hypothetical stages in the evolution of the insect form.
(Redrawn from Ross, 1965.)

Evolution of the Insects

(Carpenter - 1953)

Four stages in insect evolution

1. Appearance of primitive wingless insects
2. Development of wings
3. Development of metamorphosis
4. Development of wings flexion mechanisms

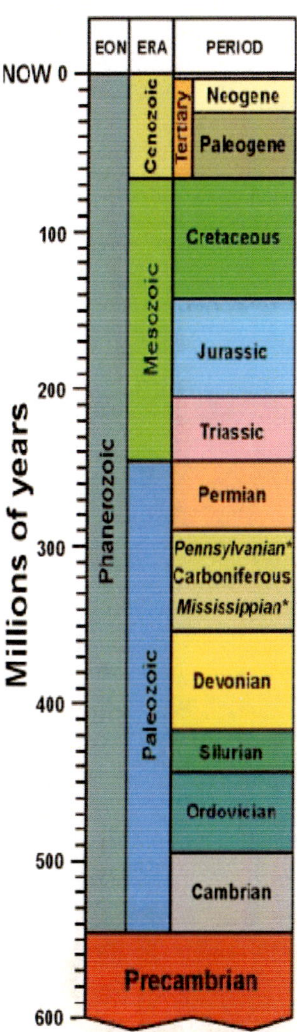

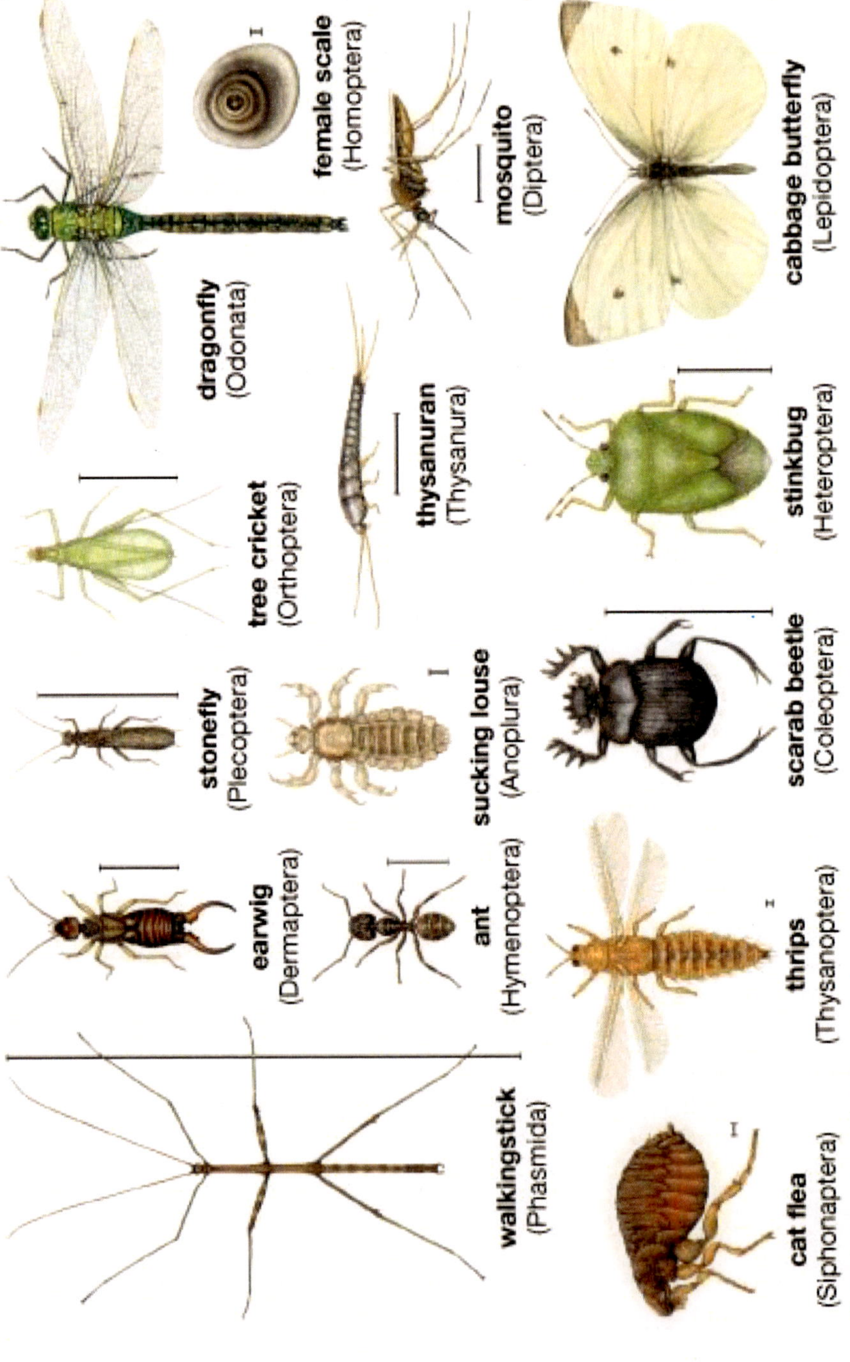
walkingstick
(Phasmida)
earwig
(Dermaptera)
stonefly
(Plecoptera)
tree cricket
(Orthoptera)
dragonfly
(Odonata)
female scale
(Homoptera)
ant
(Hymenoptera)
sucking louse
(Anoplura)
thysanuran
(Thysanura)
mosquito
(Diptera)
cat flea
(Siphonaptera)
thrips
(Thysanoptera)
scarab beetle
(Coleoptera)
stinkbug
(Heteroptera)
cabbage butterfly
(Lepidoptera)
© 2010 Encyclopædia Britannica, Inc.

Characters of Phylum Arthropoda

- Animals are triploblastic and bilaterally symmetrical.
- Body is segmented and jointed externally.
- Body is covered with hardened and nonliving chitinous exoskeleton called as chitin or body wall.
- A variable number of segments carry paired and jointed appendages.
- Animals are coelomates and the body cavity is called as haemocoel.
- The circulatory system is of open type with dorsal blood vessel carrying paired ostia.
- The central nervous system consists of a brain and ventral never cord.
- The respiration is with the help of trachea, or through the body surface or through gills or book lungs.
- The sexes are usually separate and male and female animals are easily distinguishable.
- They usually undergo metamorphosis.

The Insecta (insects) are a class of large animal Phylum called Arthropoda (arthropods)- a name that refers to the jointed limbs. The other major Classes of the living arthropods (i.e. animals related to insects)include the Crustacea (crabs, lobsters, shrimps, barnacles, woodlice etc.), the Myripoda (millipedes, centipedes etc.) and the Arachnida (scorpions, king crabs, spiders, mites, ticks etc.) and a class of extinct arthropods, the Trilobita (trilobites) known only from their fossil remains.

Phylum Arthropoda is divided into following classes:

Character	Crustacea	Arachnida	Myriapoda		Onychophora	Insecta
			Chilopoda	Diplopoda		
Habitat	Aquatic and Terrestrial	Terrestrial	Terrestrial	Terrestrial	Terrestrial	Many Terrestrial and a few aquatic.
Body regions	2 (Cephalothorax and abdomen)	3 (Pro, meso and meta soma)	2 (Head and multi-segmented trunk)	2 (Head and multi-segmented trunk)	Worm like unsegmented in adults.	3 (Head, thorax and abdomen)
Antenna	2 pairs	Absent	1 pair	1 pair	1 pair	1 pair
Visual organs	1 pair of stalked compound eyes	1 pair of simple eye	1 pair of simple eye	1 pair of simple eye	1 pair of simple eye	Compound eyes and simple eye
Locomotor organs	5 pairs of biramous appendages	4 pairs	1 pair/segment	2 pair/segment	Many bilateral lobe like legs	3 pairs of thoracic legs and 2 pairs of wings
Respiration	Gills	Book lungs and trachea	Tracheal	Tracheal	Tracheal	Tracheal
Habit	Hervibores, carnivorous	Hervibores, predators	Carnivorous	Hervibores	Organic matter	Hervibores, predators and parasitiods
Examples	Crayfish, Crabs, Woodlice, lobsters	Spiders, scorpions, ticks, mites	Centipedes	Millipedes	Peripatus	Insects

Hypothetical phyllogeny of the apterigotes and their relationship with the pterygotes (Boudreaux, 1979).

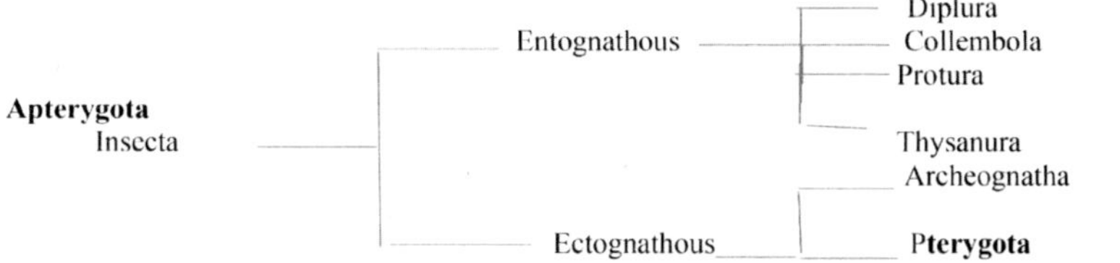

Character	Entognathous	Ectognathous
Mouth parts	Not visible externally; lie in cavity on the head produced by genae which extend ventrally as oral folds.	Visible; external to the head; not enclosed in oral pouch.

Difference between Apterygota and Pterygota

Characters	Apterygota	Pterygota
Wings	Wingless (Absent)	Winged or secondarily wingless
Members	Silverfish, springtails	Butterflies, beetles, bugs, grasshoppers
Metamorphosis	Absent/slight (Ametabola)	Partial (Hemimetabola) or complete (Holometabola)
Molting in adults	Several times	No molting in adults
Mandibular articulations	Dicondylic (Double)	Monocondylic (Single)
Mouthparts	Entognathous (mouthparts are retracted within the head i.e.hidden inside head	Ectognathous (mouth parts are exposed and well developed)
Pleural sulcus in thorax	Absent	Present
Pregenital abdominal appendages	Present	Absent
Abdominal segments	11-12	8-10
Development	Primitive	Developed
No of orders	4	26

Difference between Exopterygota and Endopterygota

Characters	Exopterygota	Endopterygota
Wings development	External (wing pads)	Internal (Wing buds)
Metamorphosis	Incomplete (Hemimetabola)	Complete (Holometabola) or partial (Paurametabola)
Life stages	3 (Egg-Nymph-Adult)	4 (Egg -larva-pupa-Adult)
Pupal stage	Absent or rarely present	Present
Immature stages	Resemble adults in structure and habit	Differ from adult in structure and habit
Immature stages	Naiad/Nymph	Larva (Caterpillar, Maggot, grub etc.)

No. of orders	16	9
Members	Bugs, grasshoppers	Butterflies, true flies, beetles and weevils etc.

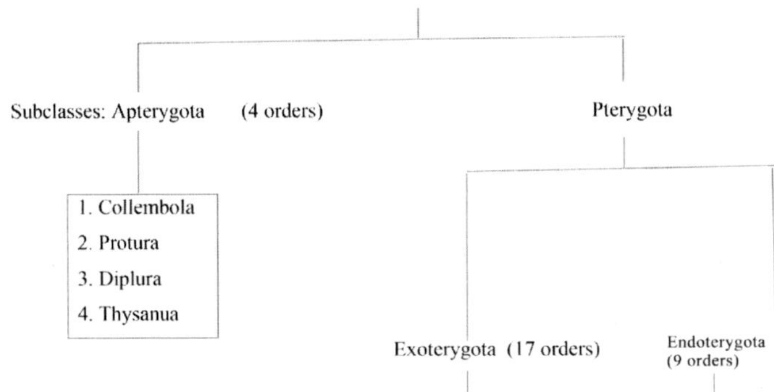

Paleopteroid	Orthopteroid	Hemipteroid	
1. Ephemeroptera	3. Plecoptera	13. Psocoptera	1. Neuroptera
2. Odonata	4. Embioptera	14. Siphunculata	2. Mecoptera
	5. Orthoptera	15. Mallophanga	3. Lepidoptera
	6. Phasmida	16. Hemiptera	4. Trichoptera
	7. Dictyoptera	17. Thysanoptera	5. Diptera
	8. Dermaptera		6. Hymenoptera
	9. Isoptera		7. Coleoptera
	10. Zoroptera		8. Siphonoptera
	11. Mantophasmatodea		9. Strepsiptera
	12. Grylloblatodea		

Panorpoid complex - Endopterygota (order 1-6)

Insect Orders (Imms Classificaton)

Order	Common Name
Apterygota Group	
1. Collembola	Springtails, snow flea
Protura (Prot-ura, from Greek protos = first, oura = tail)	Two-pronged Bristletails Proturans or Telsontails
3. Diplura	Diplurans or Japygids
4. Thysanura	Bristletails
Exopterygota Groups	
5. Ephemeroptera	Mayflies
6. Odonata	Dragonflies
7.Plecoptera	Stoneflies
8. Grylloblattodea	Rock crawlers

9.Orthoptera	Crickets, Grasshoppers and Locusts
10. Phasmida	Stick and Leaf Insects
11. Dermaptera	Earwigs
12. Embioptera	Web-spinners/Embids
13. Dictyoptera	Cockroaches and Mantids
14. Isoptera	Termites
15. Zoraptera	Zorapterans
16. Psocoptera	Psocids or Booklice
17. Mallophaga	Biting Lice/bird lice
18. Siphunculata (Anoplura)	Sucking Lice (Head and body louse)
19.Hemiptera	True Bugs
20. Thysanoptera	Thrips
21. Mantophamtodea	Gladiator
Endopterygota Groups	
22. Neuroptera	Alder Flies, Snake Flies, Aphid lion,Ant lions, Owl flies, Mantispid flies and Lacewings
23. Mecoptera	Scorpion Flies
24. Lepidoptera	Butterflies and Moths
25. Trichoptera	True Flies
26. Diptera	True Flies
27. Siphonoptera	Fleas
28. Hymenoptera	Bees, Wasps and Ants
29. Coleoptera	Beetles and weevils
30. Strepsiptera	Stylopids

The *largest insect order is coleoptera* (beetles), followed by lepidoptera (moths & butterflies), hymenoptera (ants, bees, wasps) and Hemiptera (true bugs). Hemiptera is the largest order in Hemimetabolous insects.

- **New Insect Order:** For the first time in 87 years, researchers have discovered an insect that constitutes a new order of insects. Dubbed “the gladiator” (for the recent movie), it lives in the Brandberg Mountains of Namibia, on the west coast of Southern Africa. Entomologist Oliver Zompro of the Max Planck Institute of Limnology in Plön, Germany, who identified the creature as unique, said it resembles “a cross between a stick insect, a mantid, and a grasshopper.” The discovery of the new insect order, which has been named *Mantophasmatodea*, increases the number of insect orders to 30.

Classification as Per Royal Entomological Society (Latest Classification)

Hexapoda is divided into two classes: the Entognatha includes primitively wingless hexapods such as springtails, while all the ‘true’ insects are subdivided into five major groups also know as superorders, the Apterygota, Palaeoptera, Polyneoptera, Paraneoptera and Endopterygota.

Sub class	Description	Orders	Examples
Apterygota	Primitive wingless insects with incomplete metamorphosis.	Archaeognatha (or) Microcoryphia	Bristletails
		Zygentoma	Silverfish , firebrats
Palaeoptera	Primitive winged insects, with their wings held upright or outstretched at rest and incomplete metamorphosis.	Ephemeroptera	Mayflies
		Odonata	Dragnoflies and damselflies
Polyneoptera	Winged insects, with a broad, fan-like extension to their hind wings, and incomplete metamorphosis.	Dermaptera	Earwigs
		Dictyoptera	Cockroacses, mantids , termites
		Embioptera	Embids, web spinners
		Grylloblattaria	Rock crawlers /ice crawler
		Mantophasmatodea	Heel walkers
		Orthoptera	Grasshopper, crickets
		Phasmida	Stick and leaf insects
		Plecoptera	Stoneflies
		Zoraptera	Zorapteran
Paraneoptera	Higher insects, with mostly incomplete metamorphosis, where a nymph generally resembles the adult. True bugs, lice and thrips	Hemiptera	True bugs
		Phthiraptera	Sucking and biting lice
		Psocoptera	Book lice, bark lice
		Thysanoptera	Thrips
Endopterygota	Higher insects, with a clear metamorphosis from larva via a pupa to adult, also called Holometabola. The larvae look very different to the adults, and undergo metamorphosis in a pupa where the wings develop internally	Coleoptera	Beetles, weevils
		Diptera	True flies, mosquitoes
		Hymenoptera	Bees, wasps
		Lepidoptera	Moths and butterflies
		Mecoptera	Scorpion flies
		Megaloptera	Alder flies
		Neuroptera	Lace wings and ant lions
		Siphonaptera	Flea
		Raphidioptera	Snakeflies
		Strepsiptera	Stylops
		Trichoptera	Caddisflies

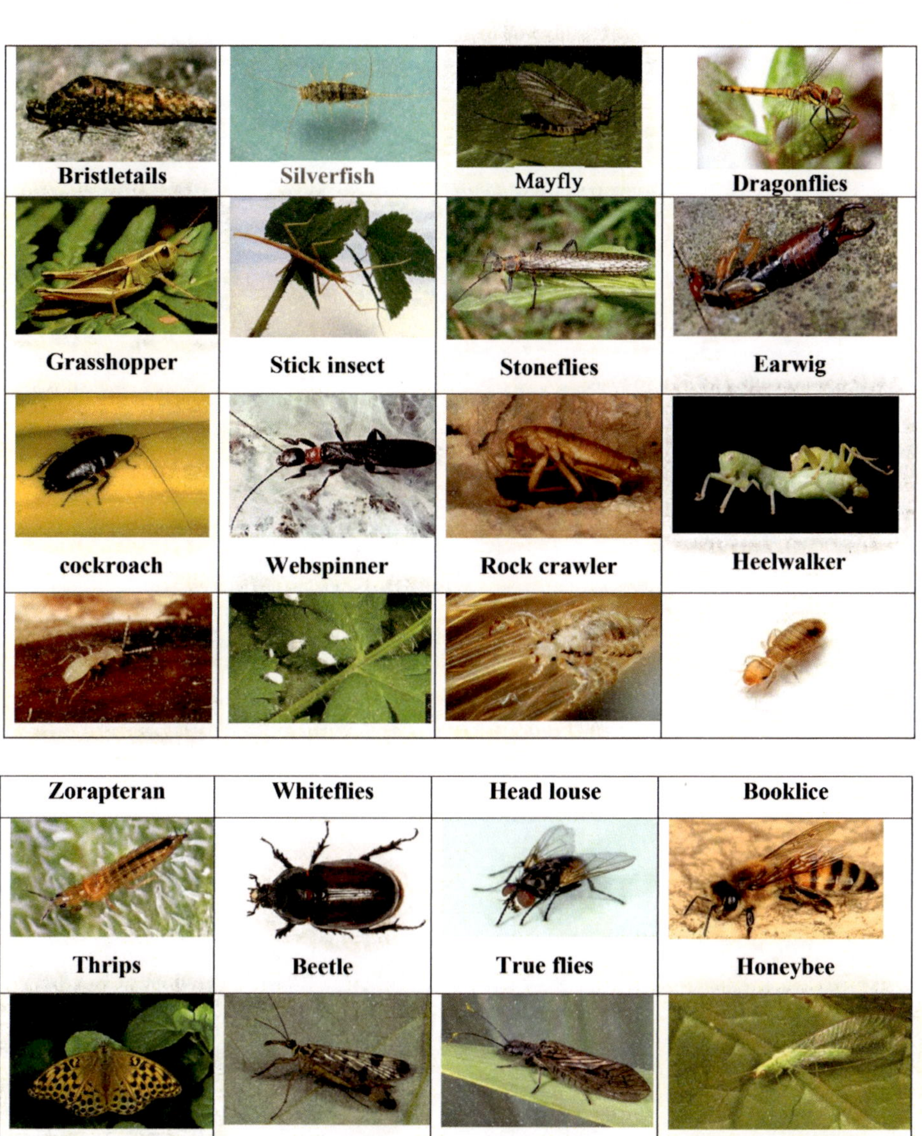
Bristletails
Silverfish
Mayfly
Dragonflies
Grasshopper
Stick insect
Stoneflies
Earwig
cockroach
Webspinner
Rock crawler
Heelwalker
Zorapteran
Whiteflies
Head louse
Booklice
Thrips
Beetle
True flies
Honeybee
Butterflies
Scorpionflies
Alderflies
Lacewings
Fleas
Snakeflies
Stylopids
Caddisflies

14

Insect Orders

Apterygotes

The Apterygota, which formerly included the other primitively wingless insects currently placed in the class Entognatha, are now restricted to the two orders Archaeognatha and Zygentoma, which in turn were formerly united as the Thysanura. Despite the superficial similarity of the two groups, it is now clear that they are not closely related, mainly because of fundamental differences in the mouthparts. The Archaeognathous mandibles are monocondylic, having a single articulating point with the head so that the mandible can rotate. The Zygentomous mandibles are dicondylic, with two articulating points that restrict the motion to a single plane yet enable the development of a much stronger biting action; this is the type found in all the higher insects, and the Zygentoma may well be the sister group of the Pterygota, or winged insects. Clearly the 'Apterygota' is not a monophyletic group and is simply retained for convenience in grouping these two orders that superficially resemble each other and have similar life histories.

Order	Remarks	Examples
Collembola (coll-glue; embol-wedge or peg) Spring tail, snow flea.	• Springtails are named for a forked jumping organ (the furcula) found on the fourth abdominal segment, retracted by a special catch (the tenaculum/hamula) on the third abdominal segment. Ventral tube (collophore/glue bar) present on first abdominal segment. • Compound eyes absent or reduced. • Antennae 4 to 6- segmented. • Abdomen 6-segmented.	*Sminthurus virdis* (the Lucerne flea) is a pest of alfalfa in Australia and *Hypogastrura armata* has been a frequent pest of commercial mushrooms.

Order	Remarks	Examples
Protura (Prot-ura, from Greek protos = first, oura = tail)	• Minute wingless insects, without eyes or antennae. • Mouthparts for piecrcing and sucking. • Front legs held forward in the manner of antennae. Tarsi 1-Segmented. • Cerci absent • Metamorphosis slight or wanting.	Proturan  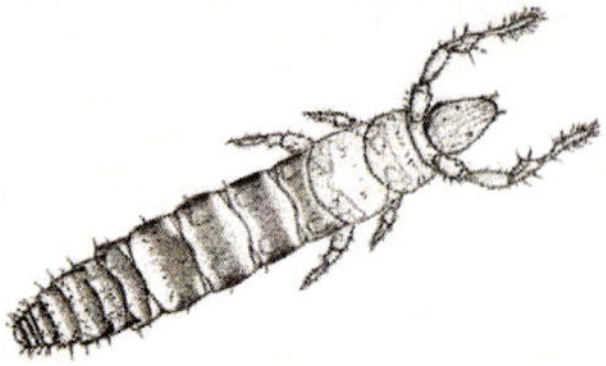
Diplura (Dipl-ura, from Greek diplos = double, oura = tail)	• Compound eyes absent. • Antennae longer than head with 10 or more bead like segments. • Abdomen with 10 visible segments • Cerci present • Tarsi 1-segmented. • Short lateral styli and eversible vesicles present on most of the first 7 abdominal segments.	Diplurans 
Thysanura (Thysan-fringe; ura- tail) Silverfish, Fire brat, Bristle tail	• Small insects with compound eyes and very long, thread like antennae. • Mouthparts for biting and chewing. • Tarsi 2-4 segmented. • Body often covered with scales. • Abdomen with long cerci and a median tail filament. These processes are fringed with bristles. • Metamorphosis is slight or wanting. • Silverfish (*Ctenolepisma sp.* and *Lepisma saccharina* feed on wallpaper paste, book bindings and the starch sizing of some textiles).	Silverfish, Fire brats  

Exopterygotes

Ephemeroptera

(Ephemero-ptera, from Greek words "*ephemeros*" meaning living a day;"*pteron*" meaning wings. **May flies**

- This is a reference to the short lifespan of most adult may flies.
- Antennae short and bristle-like

- Mouthparts in adults are atrophied.
- Incomplete development (egg, nymph, adult).
- Immatures are aquatic **(naiads)** having bilateral abdominal grills and a long cerci.
- Transparent wings, of which the forewings are larger than the hindwings. Occasionally the hindwings are absent
- Primitive wing venation.
- Unable to fold the wings over the back **(paleopteran)**
- The first winged stage is called the **subimago** a brief transitional stage that molts again into a sexually mature adult (imago).
- Mayflies have paired genital openings.
- Most species produce 500–3000 eggs, but values range from less than 100 in *Dolania* to 12,000 in *Palingenia*.
- Adults did not feed
- Three long filaments at rear of abdomen
- Naiads are important fish food.
- Bio indicators of water pollution.
- Ephemeroptera constitutes a small order of extant insects, with approximately 40 families, 440 genera, and 3330 species

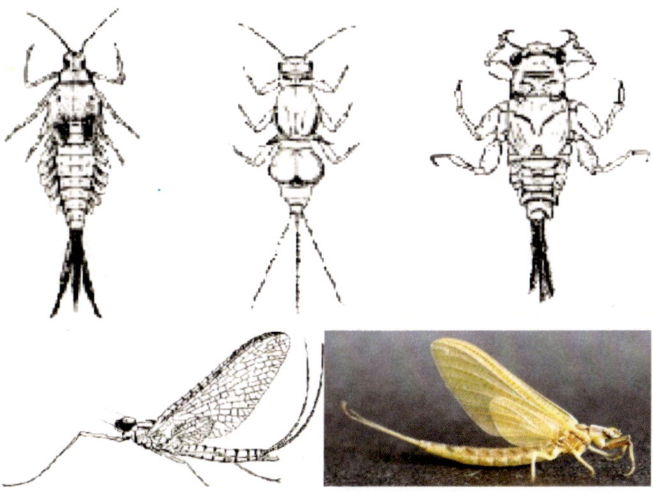

Odonata (Odon-tooth)

- Snake doctors, "devils darning needles" and mosquito hawks.
- Medium to large sized insects and attractively coloured
- Head is globular and constricted behind into a petiolate neck.
- Compound eyes are large. Three ocelli are present
- Mouthparts are adapted for biting. Mandibles are strongly toothed Lacinia and galea are fused to form mala which is also toothed.
- Wings are either equal or sub equal, membraneous; venation is net work like with many cross veins. Wings have a dark pterostigma towards the costal apex. Sub costa ends in nodus. Wing flexing mechanism is absent.

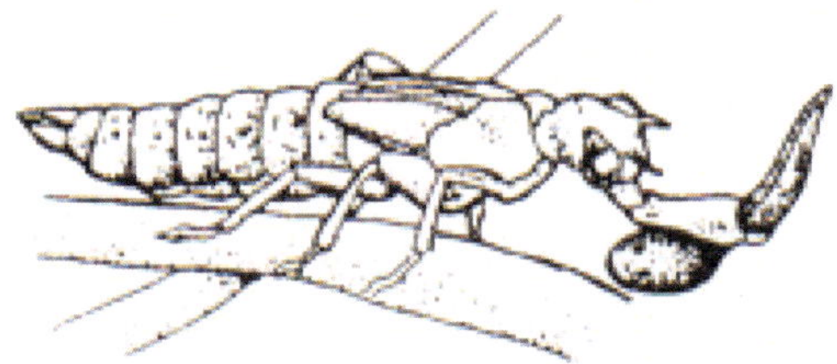

- Primitive wing structure and venation. Wings have a dark pterostigma (thickening of the wing membrane between C and R) towards the costal apex.
- Lacking the ability to fold the wings over the back.
- Adults are aerial predators. Naiads are aquatic predators. Dragonflies and damselflies can be collected with an aerial net near streams and ponds especially on a sunny day. Naiads can be collected from shallow fresh water ponds and rice fields.
- In male, gonopore is present on ninth abdominal segment. But the functional couplatory organ is present on the second abdominal sternite. Before mating sperms are transferred to the functional penis.
- Cercus is one segmented.
- Incomplete development (egg, nymph, adult).
- Immature are aquatic (**naiads**) having modified mouthparts known as the **mask** (Labuim modified).
- Dragonflies and damselflies are predaceous both as immature and adults (beneficial insects).

- There are two sub-orders. Dragonflies are classified under **Anisoptera** and damselflies are grouped under **Zygoptera.**

Classification

Odonata have two sub-orders classified in to Anisoptera (Dragonflies) and Zygoptera (damselflies).

Features	**Anisoptera (Dragonflies)**	**Zygoptera (damselflies)**
Adults		
Eyes	Meet dorsally and occupy largest part of head (**Holoptic**)	Smaller, button like (**Dichoptic**)
Sutures/sclerites	Occipit, frons and vertex distinct	Sutures are less distinct or absent
Wings at rest	Hindwings broader basally and held horizontally at rest. 	Identical and held vertically at rest. 
Wings	**different size** 	**same size** 
Wing veination	Not similar in both fore and hind wings	Similar in both fore and hind wings
Flying ability	Strong flier	Weak flier
Labium	Middle lobe deeply cleft	Variable
Male genital tract	Penis jointed: possess 2 superior anal (Cerci) and 1 inferior anal plate (Epiproct)	Penis not distinctly jointed: possess 2 superior (Cerci) and 2 inferior anal plate (paraproct).
Ovipositor	Normal or atrophied	Complete
Oviposition	Exophytic	Endophytic
Naiads		
Size	Large and robust	Small and slender

Features	Anisoptera (Dragonflies)	Zygoptera (damselflies)
Gills	Rectal gills	3 caudal gills
Propulsion	Able to propel themselves by forcibily ejecting water through anus from rectum	Lack jet propulsion mechanism

Plecoptera

(Pleco-*ptera*, from Greek *Plekein* = Plated fold, pteron = wing).

- The name Plecoptera derived from the Greek "*pleco*" meaning folded and "*ptera*" meaning wing; refers to the pleated hind wings which fold under the front wings when the insect is at rest.
- The order represent more than 3,497 species in the world.
- The insects are soft bodied having three segmented tarsi
- Filiform antennae
- Mandibulated mouthparts
- Two compound eyes, two or three ocelli,
- Two usually long cerci
- 10-segmented abdomen with vestiges of the eleventh segment.
- Adults have two pair of membranous large wings (sometimes reduced or absent), and subequal fore and hind wings (hind wings slightly wider) that fold horizontally over and around abdomen when at rest (hence the name: plecos = folded; pteros = wings)
- Nymphs are similar to adults with a closed tracheal system with or without filamentous gills. When present, gills are located on different parts of the body.
- Plecoptera nymphs are aquatic and live mainly in cold, well-oxygenated running waters, Few species are adapted to terrestrial life in Sub-Antarctic areas:

- The life cycle of **stoneflies** lasts for one or more years, but there are also bi- or tri-voltine species. Nymphal or egg diapause is not uncommon. The nymphs can moult up to 33 times before emerging.
- Nymph feed on animal or vegetable matter as collectors, scrapers, shredders or predators
- Stoneflies require clean, well oxygenated water to survive. They are extremely sensitive to water pollution and are used by ecologists as **indicators of water purity.**

Family Perlidae - Common Stoneflies (*Acroneuria mela* Frison)

Grylloblattodea

- The name Grylloblattodea derived from the Greek "*gryll*" meaning cricket and "*blatta*" meaning cockroack, refers to the blend of cricket-like and roach- like traits found in these insects.
- **Hemimetabola** -incomplete development (egg, nymph, adult)
- **Orthopteroid** - closely related to Orthoptera and Dermaptera
- Antennae slender, filiform
- Mouthparts mandibulate, hypognathous
- Body cylindrical
- Tarsi 5-segmented
- Secondarily wingless
- Cerci long, 8-segmented
- Grylloblatta sp. and rock crawlers. (1 family and 25 species worldwide)
- These omnivorous insects scavenge for food on the surface of snowfields, under rocks, or near melting ice.

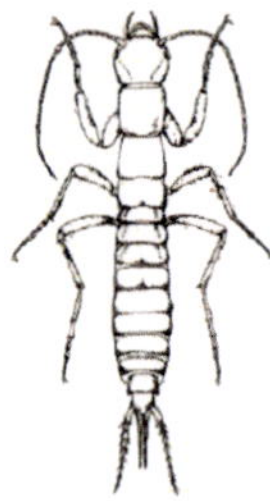

Orthoptera

(Ortho-ptera from Greek *orthos* = straight, *pteron* = wing)

- Mouthparts for biting (**mandibulate**)
- Antennae mostly **filiform**
- Prothorax large. Pronotum is curved, ventrally covering the pleural region.
- Hindlegs usually enlarged and modified for jumping (**Saltatorial**).
- Tarsi nearly always 3-4 jointed.
- Thorax generally with two pairs of wings, the front pair tough and leathery (**Tegmina**), although one or both pairs may be reduced or absent. Hindwings are membranous with large anal area. They are folded by longitudinal pleats between veins and kept beneath the tegmina.
- Cerci present, but often short and inconspicuous.
- Female with a well-developed ovipositor.
- Simple metamorphosis.
- Eggs are laid in groups as **Egg pod**.
- Stridulatorial organs are alary and femoro-alary.
- Auditory Tympanal organs present.

Classification

The order is sub divided into two suborders, viz., Caelifera and Ensifera

Sub Order: Ensifera (Long-horned Grasshoppers)	**Caelifera (Short horned Grasshopper)**
Antenna longer with > 30 segments.	Antenna shorter with < 30 segments.

Tympanal organ present on the fore tibia. 	Tympanal organ on the either side of the 1st abdominal segment. 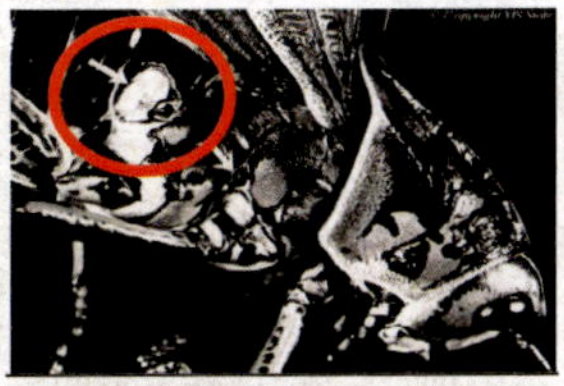
Nocturnal	Diurnal Eggs are laid in groups in soil inside shallow burrows.
Mandibles are specialized for feed on monocot plants.	Mandibles are specialized foe feed on dicot plants.
Tactile responses is well developed.	Vision and hearing acute.
Rely on crypsis	Rely on jumping to escape from predators.
Walking type legs	Jumping type legs
Arboreal	Terrestrial (not arboreal)
Eggs are laid in **groups** in soil inside shallow burrows (**Egg pod**)	Eggs are **singly** inserted into plant tissue or soil
Family:	
Gryllidae- House Cricket (*Gryllus sp.*)	**Acridiidae**-Locusts and short horned grasshopper
Tettigoniidae- Green Katydid (*Holochlora indica*) Surface grasshopper – *Conocephalus indicus*	Rice grasshopper- *Hieroglyphus banian* Small Rice Grasshopper-*Oxya velox* Cotton grass hopper - *Cyrtacanthacris ranacea* Calotropis grasshopper - *Poecilocerus pictus*
Gryllotalpidae-.Mole cricket (*Gryllotalpa gryllotalpa*)	

Sub order: Caelifera

Class - Acrididae: (Locusts, Grasshoppers)

- Antenna is short
- Tarsus is three segmented
- Ovipositor is short and horny
- Tympanum is located one on either side of the first abdominal segment.
- Sound is produced by femoro-alary mechanism. A row of peg like projections found on the inner side of each hind femur is rubbed against the hard radial vein of the closed tegmen.
- Locusts are a serious threat to tropical agriculture. They swarm under favourable conditions and mainly feed on grasses, cereals etc.
- Some species have fairly elaborate courtship. Mating itself may take up to one hour, and male may ride on back of female for a period of a day or more, a behavior known as mate guarding. Females oviposit in loose soil (typically), among plant roots, in rotting wood, or even in dung. Clutches consist of 10-60 eggs, and females may lay up to 25 clutches over several weeks. Oviposition typically occurs in late summer, and the egg (as a developing embryo) overwinters. Eggs then hatch in spring. Life cycle is typically one year. A few species overwinter as juveniles (nymphs).

Sub order : Ensifera

Class - Tettigonidae : (Katydids, Long horned grasshoppers)

- Antenna is long, slender as long as or longer than the body.
- Tarsus is four segmented.
- Ovipositor is sword like.
- Auditory organs are found in foretibiae. In each foretibia a pair of tympanum is present. The outer tympanum is larger than the inner. Sound production

is alary type. A thick region on the hind margin of the forewing (scraper) is rubbed against a row of teeth on the stridulatory vein (file) present on the ventral side of another forewing which throws the resonant area on the wing (mirrors) into vibrations to produce sound.

Gryllidae (Cricket)

- Antenna is long.
- Tarsus is four segmented.
- Ovipositor is slender and needle like.
- Forewings are abruptly bent down to cover the sides of the body.
- Hindwings are acuminate. They are produced into a pair of long processes which project beyond the abdomen.
- Cerci are long and unsegmented
- Auditory organs and stridulatory organs are similar to long horned grasshopper. Males stridulate during night. They produce a shrill chirping noise.
- Gryllus sp. It is household pest.

Gryllotalpidae : (Mole crickets)

- They are brown coloured insects found inside the burrows. Eyes are reduced.
- Pronotum is elongate, ovate and rounded posteriorly.
- Forelegs are fossorial. Tibiae are expanded and digitate.
- Hindwings are extended beyond the tegima as a pair of processes

- Special stridulatory structures are absent. A humming sound is produced by rubbing the forewings.
- A pair of tympanum is found on the under surface of the tibiae.
- Ovipositor is vestigial.
- Mole crickets burrow into the soil and feed on tender roots of growing plants. *Gryllotalpa africana* is a pest on stored potatoes.

Locusts

- Locusts are grasshoppers which multiply enormously under certain favourable conditions, forms swarms and migrate in millions from place to place or country to country feeding on all vegetation on their way. There is a kind of an **Ecological opportunism** resulting in a group effect. Uvarov proposed 2 phase theory:
- 2 phases in locusts differing structurally
    - The *Phasis solitoria* (Solitary phase)
    - The *Phasis gregaria* (Gregarious phase)

Features	***Phasis solitoria* (Solitary phase)**	***Phasis gregaria* (Gregarious phase)**
Nymphs		
Colour	Green, grey or brown	Black, yellow or orange
Behaviour	Solitary feeder	Bands/group feeders
Number of molts	5-6	5
Adults		
Colour	Greenish/Grey throughout	Pink (immature), Yellow (mature)
Behaviour	Night flier, singly	Day flier in swarms
Pronotum	Longer and crested Pronotum.	Shorter and saddle shaped Pronotum.
Antennae	27-30 segmented	27 segmented
Eye stripes	6-7	6

Features	***Phasis solitoria*** **(Solitary phase)**	***Phasis gregaria*** **(Gregarious phase)**
Femoral spines	Both weak and strong; longer hind femur.	Only weak; shorter hind femur.
E/F ratio	2.06 or below.	2.16 or above.
F/C ratio	Male: 3.75 or above Female: 3.85 or above.	Male: 3.15 or below. Female: 3.85 or below.

Locust Species

Y.R. Rao from India has authority on Grasshopper.

• Bombay locust	:	*Patanga succinata*
• Desert locust	:	*Schistocerca gregaria*
• Migratory locust	:	*Locusta migratoria (most destructive)*
• Moroccan locust	:	*Dociostaurus moroccanus*
• African Red locust	:	*Nomadacris septemfasciata*

Management of Locusts and Grasshoppers

- Deep ploughing, digging of harrowing of land to destroy egg pods.
- Use of flame throwers to burn gregarious hoppers.
- Trenching (45 cm deep and 30 cm wide)
- Trimming of bunds.
- Poison baits.
- Dusting with chlorpyriphos (2%).
- Spraying of crops with chlorpyriphos (2 ml/litre of water).

Locust Warning Organization was established in 1939 with Headquarter at Jodhpur (Rajasthan). The Organization is working under the control of Plant Protection Advisor to the Government of India, Directorate of Plant Protection, Quarantine and Storage (DPPOQ&S), Faridabad.

Phasmida

(*Phasmida* from Greek *phasma* = an apparition) refers to the cryptic appearance and behaviour of these insects.

- Large, apterous or winged insects, frequently of elongate, cylindrical form, more rarely depressed and leaf-like.
- Antennae long, slender.
- Mouthparts mandibulate.
- Prothorax short; mesoand metathorax usually elongate, the latter closely associated with 1st abdominal segment.
- Phasmid eggs often resemble seeds. The eggs may remain dormant for over a year before hatching.
- Legs similar to each other; coxae small and rather widely separated; tarsi almost always 5-segmented.
- Fore wings, when present, usually small and with submarginal costa. Wing-pads do not undergo reversal during development.
- Ovipositor small and mostly concealed by enlarged 8th abdominal sternum. Male external genitalia variable and asymmetrical, concealed by 9th abdominal segment.
- Cerci short, unsegmented. Specialized auditory and stridulatory organs absent.
- Eggs deposited singly.
- Metamorphosis slight.
- When attacked by a predator, the legs of some phasmids may separate from the body (**autotomy**). Some species can even regenerate lost legs at the next molt. These are the only insects able to regenerate body parts. They have abscission suture between femur and trochanter.
- Some phasmids change colour with changes in temperature, humidity or light intensity. Pigment granules in the epidermis disperse at night or on cool days, darkening the cuticle and absorbing more heat.
- "Gladiator' was chosen as a common name for these insects because of their physical similarity to armoured fighters in the movie Gladiator.
- Leaf insects (*Phyllum sp.*) and stick insects (*Carausius sp.*).
- Australian tick insect *Megaphasma australicus* is the longest in the world measuring about 33 cm excluding the antennae.

Carausius morosus *Phyllium bilbatum*

Dermaptera (Derma-skin; ptera-wing; Earwigs)

- The name Dermaptera derived from the Greek "derma" meaning skin and "ptera" meaning wings refers to the thickened forewings that cover and protect the hind wings.
- Front wings short and leathery (Tegmina).
- Hind wings semicircular and pleated.
- Body elongated, dorso-ventrally flattened
- Prognathous head
- Ocelli absent in some; compound eyes in most species but reduced or absent in some
- Segmented antennae,
- Biting and chewing mouthparts.
- Cerci enlarged to form pincers (forceps like)
- Paired penes or penis present.
- Females lay their eggs in the soil and may guard them until they hatch (Parental care).
- **Two sub orders**

Arixeniina	**Hemimerina**
two genera, *Arixenia* and *Xeniaria*, with a total of five species	Hemimerina, they are blind and wingless, with filiform segmented cerci. Hemimerina are viviparous ectoparasites, inhabiting the fur of African rodents in either. Hemimerina also has two genera, *Hemimerus Araeomerus*

- The European earwig, *Forficula auricularia* cause economic losses in fruit and vegetable crops.

- *Euborellia annulipes* common groundnut earwig.

Embioptera (Embia-lively; ptera-wings; Webspinners/Embiids)

- The name Embioptera derived from the Greek "embio" meaning lively and "ptera" meaning wings refers to the fluttery movement of wings that was observed in the first male Embioptera described.
- small to medium insects, ranging from 0.06 to 0.78 inches (1.5 to 20 millimeters) in length. They have long, narrow bodies that are usually brown or black in color.
- Antenna is filiform.
- Mouthparts mandibulate, prognathous
- Only adult males have wings.
- Tarsi 3-segmented.
- Hind femur enlarged, adapted for **running backward**
- The tarsi of the front legs are enlarged and contain glands (about 100 no.) that produce silk used to construct elaborate nests and tunnels under leaves or bark.
- Wings present only in adult males, highly flexible, smoky in color
- Cerci 1-2 segmented; asymmetrical in male
- The abdomen ends in a pair of short, fingerlike projections bristling with tiny hairs. These structures function like antennae and help to guide their backward movement within the galleries.
- Maternal care exhibited by females.
- On starvation both sexes become cannibalistic.

- Species -*Metoligotoma pentanesiana* (Australembiidae), *Archembia* n. sp. (Archembiidae), *Haploembia tarsalis* (Oligotomidae), *Macrembia* sp. ('Embiidae'), and *Antipaluria urichi* (Clothodidae)

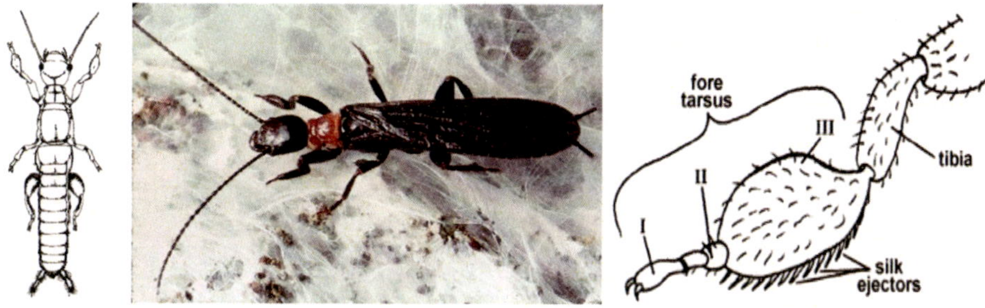

Silk glands

Silk covering of different Embids

Dictyoptera (*Diction*-network, *pteron*-wing)

- Large to medium sized insects
- Body dorso-ventrally flattened, depressed, oval shape (cockroaches) or very long bodied, stick like in some (mantids)
- Usually *Hypognathous*
- Antennae long and filiform.
- Mouthparts for biting and chewing type.
- Usually two pairs of wings of which the front ones are leathery and held flat over the back, slightly overlapping in the midline. *Hind wings are membranous* with a large anal lobe folded in a fan like fashion, often fan like. In many species, wings are reduced or absent altogether.

- Legs long and spiny, and all fairly similar (Cockroaches) or the fore legs raptorial (Mantids). All three pairs are similar in Blattidae (cockroaches) useful for running (ambulatory), while in Mantidae (praying mantids), forelegs are modified for grasping the prey(Raptorial), and middle and hind legs are normal used for running.
- Tarsi 5-segmented.
- Abdomen with conspicuous, segmented cerci.
- Genitalia of both sexes reduced, concealed behind 7th abdominal segment in female, and behind 9th abdominal segment in males.
- Metamorphosis simple with 5-12 nymphal moults according to species.
- Eggs are produced in batches contained in a hard, protective capsule called the ootheca (**Egg case/Shepherd's purse**) formed by the secretions of Collateral glands
- Terrestrial, tropical and subtropical insects. The order Dictyoptera contains nearly 6,000 species, distributed worldwide. With very few exceptions, mantids and cockroaches require terrestrial habitats.
- Cockroaches are *omnivorous*, while Mantids are *predacious* on other insects, small animals
- Most species live in tropical regions.

Classification

Dictyoptera is divided into two suborders viz., Blattaria (cockroaches) and Mantodea (preying mantids). There are two important families viz., Blattidae and Mantidae

Blattidae	Mantidae
• The cockroaches, often known as "waterbugs" are scavengers or omnivores. • Head not mobile in all directions; hidden by pronotum (shield like) • *Body dorsoventrally flattened, depressed and oval shaped, Heavily pigmented body* • ***Ocelli called fenestrae*** • Antennae slender, filiform, setaceous • Pronotum oval, shield like, covering much of head and thorax. • Legs adapted for running; tarsi 5-segmented. • Gizzard armed with chitinous teeth to grind food • Female don't devour male during mating • Cerci short, multi-segmented. • Nymph not cannibalistic • No mimicry • Two pairs of wings, both are functional, used for flying. *Forewings modified as tegmina, hind wings membranous* with large anal area. • Eggs are laid inside a chitinous ootheca • Eggs are arranged in double row and are enclosed in sac called ***ootheca*** which is formed by the secretions (polystyrene) of collateral glands (modified accessory glands of females) • They are *omnivorous scavengers and will eat almost anything* (sometimes even cannibalistic), and *active mainly at night*. They are considered *pests of stored produce* because, although they eat only a little, they *contaminate large quantities* which take on a characteristic smell and have to be thrown away. Also they can be carriers of diseases spread by viruses and bacteria. Oriental cockroach (*Blatta orientalis*) American cockroach (*Periplaneta Americana*) German cockroach (*Blattella germanica*)	• Head not covered with pronotum. • Head triangular with well-developed compound eyes. • Head mobile in all directions; pronotum (elongate) • Antenna filiform, long, many segmented. • Three ocelli present • Mouthparts mandibulate, hypogonathous. • Forelegs raptorial type. • Tarsi 5-segmented. • Cerci short, multi-segmented. • No specialised stridulatory organs are present though *some Mantids do have a single ear on the metathorax* which allows them to hear the sonar of bats. • Chitinous teeth absent in Gizzard. • Female don't devour male during mating • Nymph not cannibalistic • Nymph mimic leaves and flowers • Eggs are laid in a water tight *egg case/ ootheca which is fixed to plants*. The case is formed from a frothy gum (polystyrene) secreted by collateral glands of females. After exposure to air, the case hardens. • Eggs are enclosed in filled solidfied foam. Ootheca is not chitinous • *Nymphs emerging from ootheca resemble ants.* • Generally consider to be highly beneficial insects because they feed on other insects (**Predators**). Since they are **cannibalistic** and also feed on other beneficial insects, their value as biological agents is probably rather limited. • Mantids are the only insects that can turn their head from side to side without moving any other part of the body. • A female mantid may eat her mate while he is still linked with her in copulation. • *Mantis religiosa, Gongylus gongyloides*.

American cockroach

Oriental cockroach

Oriental cockroach

Mantis religiosa

Isoptera

Termites/White Ants

- The name Isoptera derived from the Greek "Iso" meaning equal and "ptera" meaning wings; refers to the similar size, shape and venation of the four wings.
- Social, polymorphic living in communities composed of reproductive castes (kings and queen), supplementary castes and sterile castes (soldiers and workers).
- Mandibulate
- Head is movably articulated, Prognathous
- Antenna moniliform.
- Wings are of three types: Apterous, Brachypterous and Winged.
- Wings are present only in sexually mature forms during swarming season. In winged forms, the front and hind wings are similar in size, shape and venation, hence is called Isoptera. When at rest, wings are held flat over the body and extend beyond the tip of the abdomen. Wings are membranous

with somewhat reduced venation and shed by means of basal fractures along basal suture leaving wing scale/stub.

- Wings similar have a Humeral suture where the wings break off and shed. In Mastotermes, humeral suture is absent and the wings become irregularly torn off.
- Eyes well developed in winged (alate) forms, reduced or absent in others (sterile workers/secondary reproductive).
- Pro-thorax is freely movable, narrower than head. Meso and metathorax wider than head
- **Frontal gland** is a characteristic termite organ, which is greatly developed in soldiers for defense function. It is formed by the hypodermal cells in the head, is sac like gland opens in a shallow depression on the surface of the head where the cuticle is pale, is known as **fontonelle**.
- Short and stout, Tarsi usually 4 segmented
- Cerci short or very short.
- Simple or incomplete (Hemimetabolous), with three stages; egg, nymph and adult.
- Polymorphic - 3 types of castes.

Termite castes

- **Reproductive castes (Kings and queens)**
  - Winged, compound eyes and ocelli. Body may be darkly pigmented.
  - Head well developed, with chewing mouthparts moniliform (beaded) antennae.
  - Compound eyes present.
  - Two pairs of membranous wings, all similar in shape and size; wings are shed after mating.
  - The castes come out during rainy season in large numbers, mate and shed their wings. The remnants are called **Stubs**
  - The queen lives for many years
  - After fertilization, no change in queen's form except in **Termitidae**-the queen attains enormous obesity of abdomen due to increase in size of ovaries and fat bodies, incapable of movement. The condition is known as **physogastry**.

- Queen lay millions of egg (an egg/second) and live for many years. When fully engorged, she may be 14 cm long, 3.5 cm in diameter and capable of producing up to 30000 eggs per day.

- **Supplementary castes**
  - Apeterous, brachypterous having reduced eyes present in the colonies and replace reproductive caste when she later dies.
  - Production of these castes is inhibited by pheromones produced by the queen.
  - They sometimes carry on extensive reproduction in the colony and may supplement the queen in building the colony.

Sterile castes (soldiers and workers)

- Body pale in colour, somewhat ant-like in appearance but with a broader junction between thorax and abdomen.
- Compound eyes small or absent.
- Head large and cylindrical or small and round.
- Antennae beaded.
- Mouthparts chewing; sometimes with large mandibles

- **Soldiers**
  - These are also sterile wingless adults with greatly enlarged head and mandibles. Slightly larger than workers, may or may not have compound eyes. They protect the colony.
  - Two types of soldiers may be seen. (1) Mandibulate type (2) Nasute type (Nasuti).
  - In **nasute type**, the individuals have the head prolonged anteriorly into a narrow snout through which a sticky secretion is exuded. Mandibles are reduced in these. They have well developed frontal gland.

- **Workers**
  - These are sterile wingless adults.
  - They are pale in colour, lack compound eyes.
  - Mandibles relatively small. They collect food and feed queens, soldiers and newly hatched young ones.
  - They build up nests, passage ways, tunnels and galleries.
  - They may be dimorphic including major and minor forms.
  - They form the bulk (90%) of the colony.

Immature	**Adults**
Body pale in color, somewhat ant-like in appearance but with a broader junction between thorax and abdomen	Body may be darkly pigmented
Compound eyes small or absent	Compound eyes present
Head large and cylindrical or small and Round	Head well-developed, with chewing mouthparts and beaded antennae
Antennae beaded	Two pairs of membranous wings, all similar in shape and size; wings are shed after mating

Soldiers	**Workers**
Large, well sclerotized.	Poorly sclerotized.
More or less pigmented.	Non-pigmented.
Mandibles are prominent.	Reduced eyes, small brain.
Frontal glands on frons, defensive in function.	Role: gnawing of wood, attend eggs and young ones, forage food, construct nest.
Di/polymorphic	In *Kalotermes*: No fixed worker caste and the functions of workers are performed by fully grown nymphs known as **Pseudergate/false workers**.
Nasutes are the soldiers having medium rostrum and vestigial mandibles but have well developed frontal glands.	

Queen (Physogastry)

Workers

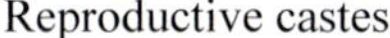

Reproductive castes

Termitarium

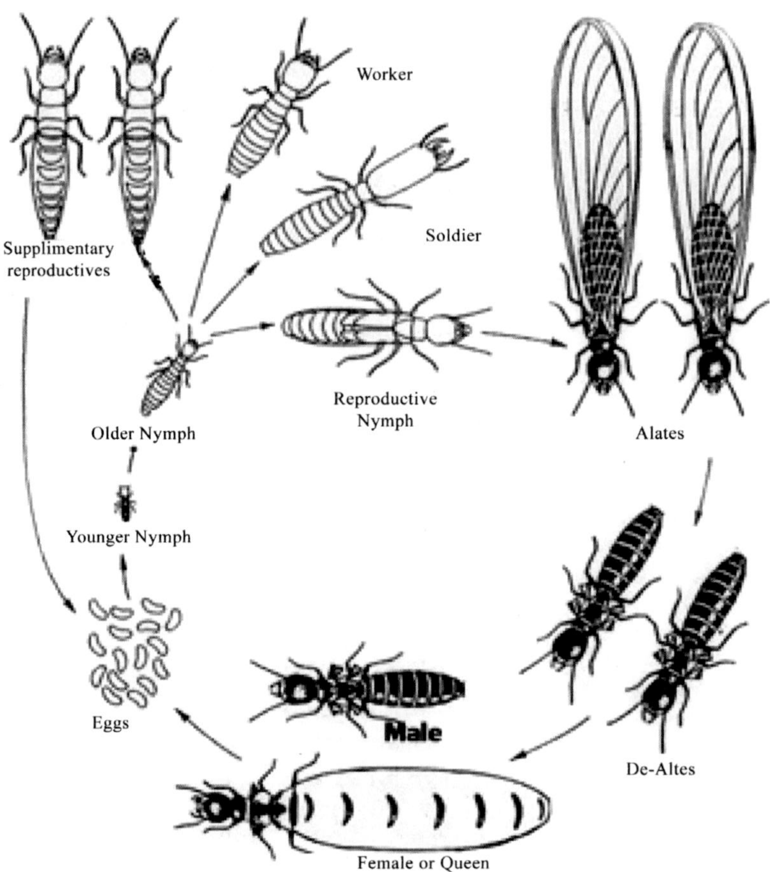

Life cycle of termites

- **Nest/Mound/Termitaria/Termitorium**
  - Height up to 20-25 feet tall and 10-12 feet diameter.
  - Very strong made of mud particles, saliva and faecal cartons.
  - Have a large network of chambers and passages.
  - Royal cell is located deep in the chamber.
  - Odontotermes construct fungus comb/garden.
  - Kings and queens are fed by workers (Trophollaxis) with regurgitated food and saliva.
  - In wood feeding termites, the digestive tract contains Trichonymphid Mastigophoran Protoza (TMP) which help in cellulose digestion.

Economic importance

- Highly polyphagous, are well known both for their destruction of human property and cause damage to crops. Feed on dead plant materials, generally in the form of dead wood, leaf litter etc.
- Economically significant as pests that can cause serious structural damage to buildings, crops or plantation forests.
- Termites are usually the most dominant organisms in tropical forest environments. Their populations typically range from 2000 to 4000 individuals per square metre but may occasionally run as high as 10, 000 individuals per square metre. A single nest may house nearly a million workers. Their biomass (up to 22 g/sq.m) exceeds the combined biomass of all vertebrate species living in the same area.
- The food of termites is the cast skins and faeces of other individuals, dead individuals and plant materials such as wood and wood products. Termites frequently groom each other with their mouth parts as a result of the attraction of some secretions available on the body (trophallaxis – Mutual exchange of food i.e secretions on the body).
- Some of the most advanced species are the Macrotermitidae which grow fungi (Fungus Garden) for food (Termitomyces) inside their nests on piles of faecal pellets. The sterile workers live for 2-4 years while primary sexuals live for at least 20 and perhaps 50 years.
- The African termite *Macrotermes bellicosus* reaches a length of 5 inches at maturity and is the world's largest termite.
- Termites are an important part of the community of decomposers. Termites become economic pests when their appetite for wood and wood products extends to human homes, building materials, forests and other commercial products.

Termite damage symptoms

Major families

- **Mastotermitidae (Primitive fossil):** *Mastotermes darwiniensis.*
- **Rhinotermitidae (Subterranean termites):** Formosan subterranean termite (*Coptotermes formosanus*).
- **Hodotermitidae (Rotten wood termites):** Pacific dampwood termite, *Zootermopsis angusticollis.*
- **Kalotermitidae (Drywood and dampwood termites)-**Without workers, *Pseudergates*, *Kalotermes sp.*
- **Termitidae-** Higher termites, ground dwelling. Odontotermes, *Microtermes sp*.

Family: Termitidae

- Included about 2/3rd of recent isopterans
- Ground dwelling with wide range of food habits
- Well developed caste system
- Mouth parts : biting types, well developed mandibles
- Fontanelle present
- Pronotum of workers is narrow with a raised anterior lobe, saddle shaped

- The scale or stub of the front wing is shorter than pronotum. Wing margin is more or less hairy. R_S reduced / absent.
- Cerci with 1 or 2 segmented
- Apterous, Brachypterous and Winged forms present
- These are all ground dwelling with wide range of food habits.
- Eg: *Odontotermes obesus, Microtermes obesi*

15

Zoraptera

- The name Zoraptera, derived from the Greek "*zor*" meaning pure and "*aptera*" meaning wingless was given to the order before winged forms were discopvered.
- Members of the order Zoraptera are small (less than 4 mm) and usually found in rotting wood, under bark or in piles of old sawdust.
- They live in small aggregations (gregarious) and appear to scavenge on spores and mycelium of fungi, or occasionally, on mites and other small arthropods.
- polymorphism in populations; with blind, apterous individuals dominating during colony life, but populations produce eyed alates (individuals with wings) for dispersal and founding new colonies.
- males with a distinctive "mating hook" and unique mating behaviors.
- unique grooming behaviors

Adults

- Antennae 9-segmented (moniliform).
- Mouthparts mandibulate, hypognathous.
- Soft bodied, small (usually less than 3 mm).
- Wings often absent with reduced venation when present.
- Tarsi 2-segmented
- Cerci 1-segmented
- ovipositor absent
- incomplete development (egg, nymph, adult)

Immatures

- Structurally similar to adults
- Always wingless

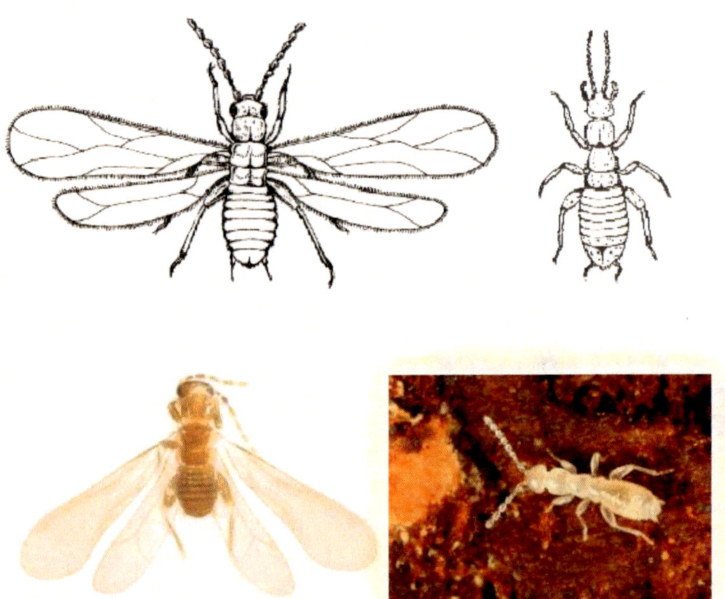

Psocoptera (Psocids/Barklice/Booklice)

The name Psocoptera is derived from the Greek "*psokos*" meaning rubbed or gnawed and "*ptera*" meaning wings; winged insects that gnaw. Also known as Corrodentia

Pearman's organ, a sound producing structure in the hind coxae of some male barklice, produces a clicking sound that attracts females. The male then performs a courtship dance in an effort to intimate copulation.

Unlike blood-sucking lice, psocids (pronounces so'-sid) are phytophagous, feeding on organic matter including algae, lichen, fungi, pollen, decaying plant particles (detritus), and occasionally dead animal matter. The habit of psocids includes living or dead foliage, ground litter, bark of trees, and inside human habitations.

- Under 2 mm in length.
- Body pale, often unpigmented.
- Head prominent with thread like antennae (15-20 segmented).
- Compound eyes large
- Narrow neck between head and thorax.

- The mouthparts of psocids are arranged in a mortal-and-pestle design, allowing the insects to scrape food from organic matter and grind it up
- Two pairs of wings (barklice); some species are wingless (booklice).
- Front wings larger than hind wings; venation reduced.
- Wings held tent-like over the body.
- Tarsi 2 or 3-segmented.
- Cerci are absent.
- Some psocids have the ability to spin silk. Dorsal pair of labial glands are modified into silk glands.
- Three suborders (Trogiomorpha, Troctomorpha, and Psocomorpha) which are distinguished by the number of segments in the antennae, tarsi, and labial palps.
- The common book louse is *Liposcelis sp.*

Barklice	**Booklice**
• Head prominent, with thread-like antennae • Ocelli present, 3 in number • Narrow "neck" between head and thorax • Two pairs of wings; some species are wingless • Front wings larger than hind wings; venation reduced • Wings held tent-like over the body • Tarsi 2- or 3-segmented	• Head prominent, with thread-like antennae • No ocelli • Narrow neck between head and thorax • Always wingless • Under 2 mm in length • Body pale, often unpigmented
• Barklice are outdoor, winged forms living on tree trunks, branches, and leaves.	• Booklice are indoor, wingless forms that are sometimes found in old books. Booklice occasionally damage books by feeding on starchy materials in the binding. They will also feed on, and often destroy, collections of dried insect specimens.

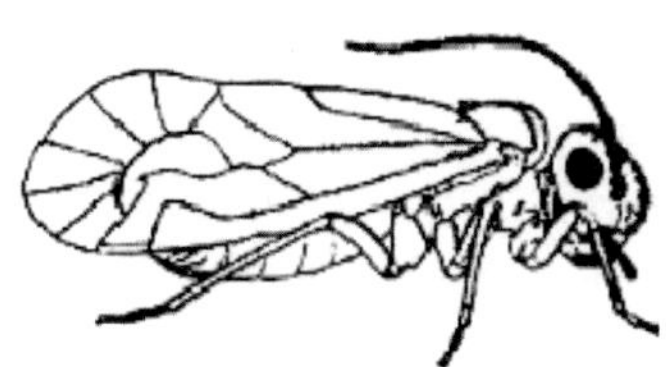

Mallophaga-Biting Lice

(Mallo-*phaga*,from Greek *mallos*=kiefer/hair,phagein=eat)

- It is also considered as suborder under order **Phthiraptera**
- Morphologically they can be easily differentiated from bloodsucking lice by the fact that their head is broader than the breast (thorax)
- Small apterous insects living as external obligate parasites of birds or, less frequently, of mammals.
- Body flattened
- Head broad,with modified mandibulate mouthparts
- Antennae short; 3-5 segmented (filiform)
- Eyes reduced or absent or have 1-2 Ommatidia
- Tarsi 1- or 2-segmented ,most species have two small Claws(Scansorial type legs)
- Body usually flattened, with prothorax distinct from the other two thoracic segments, which may be partly fused together.
- Cerci absent.
- All developmental stages stay on their hosts permanently; host-to-host transmission occurs by body contact.
- Females can lay 50 - 100 eggs which are cemented to the feathers or hair of the host. Nymphs and adults look alike. From egg to mature adult takes 3 - 4 weeks.
- They can only survive for a maximun of three days after their host has died and may hitch a ride on a passing fly (phoresis) in the hope of reaching a new host, they may also use phoresis in order to spread to a new host even if the present one is still alive

- Biting lice feed mainly on particles of skin, feathers and fur.

  Pigeon pea louse: *Columbicola columbae*

  Sparrow louse: *Bruleia subtillis*

  Domestic fowl: *Menopon gallinae*

  Domestic chicken louse *: Gallus gallus domesticus*

  Cattle Louse: *Trichodectis bovis*

  Dog louse: *T. canis*

  Cat louse - *Felicola subrostratus*

- Lice associated with domestic animals have also been implicated in the transmission of disease(e.g. hog lice, spread pox virus and cattle lice spread rickettsial anaplasmosis). Biting lice do not usually spread disease pathogens, but heavy infestations in poultry can cause severe skin irritation, weight loss, and reduced egg production.

- Lice that feed exclusively on blood do not get a ball balanced diet. To compensate for the absence of certain vitamins and amino acids these lice have intestinal symbionts (mostly bacteria) that provide additional nutrients.

- Their feeding (on keratin or blood) always requires the aid of endosymbionts in mycetomes (which are transmitted to progeny).

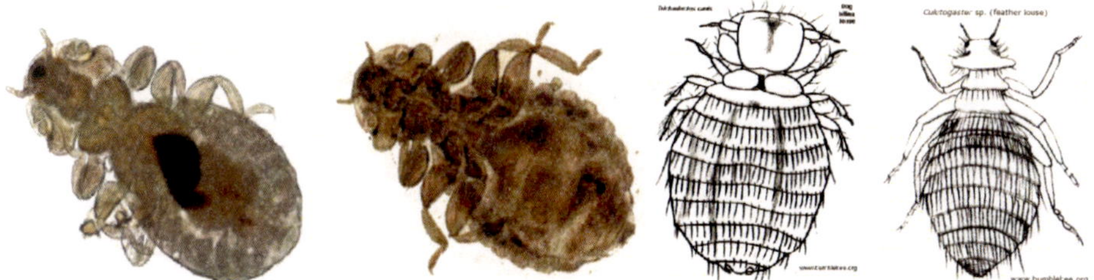

Bird lice

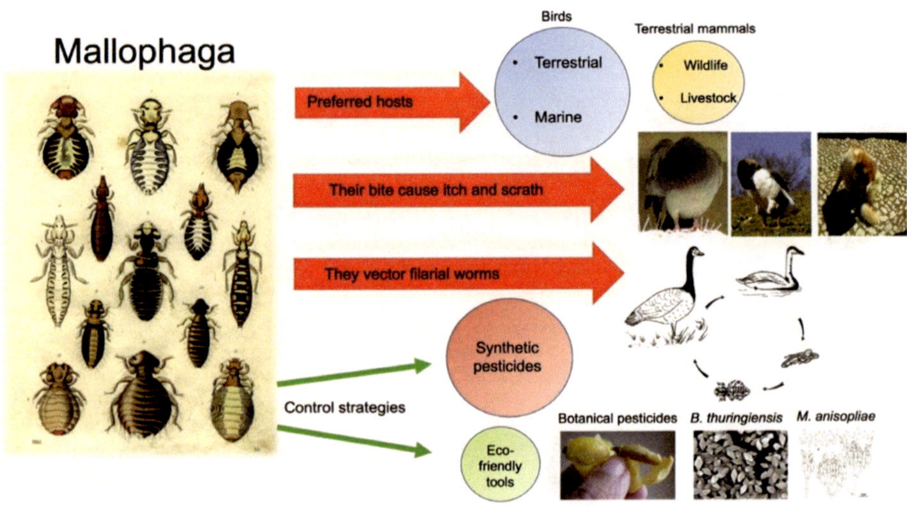

Biting Lice	Sucking Lice
• Head broad, with mandibulate mouthparts • Antennae short; 3-5 segmented • Eyes reduced or absent • Tarsi 1- or 2-segmented, most species have two small claws	• Head conical, with suctorial mouthparts • Antennae short, 3- to 5-segmented • Eyes reduced or absent • Tarsi usually 1-segmented with a single large claw

Siphunculata (Anoplura)

(Sipho-tube ; cula- a little tube; Sucking lice)

- The Anoplura are sometimes combined with the mallophaga into one order called the Phthiraptera
- Apterous, ectoparasitic, tough skinned, dorsoventrally flattened.
- Small, measuring only 0.5—5 mm in length, and oval in shape.
- Head is narrow, a distinguishing character relative the chewing lice.
- Piercing and sucking type mouth parts , retracted into head when not feeding.
- Antennae short, 3-5 segmented (moniliform).
- Eyes reduced or absent.
- Tarsi usually 1-segmented with a single large claw.
- Legs clasping/grasping type.

- Crab louse of humans: *Phthirus pubis* (*Pthiriasis* causes intense itching).
- Body louse: *Pediculus humanus corporis* (vector of relapsing/typhus/ trench fever transmitted by *Spirochaete recurrentis*).
- Human head louse (cootie): *P. humanus capitis.*
- Dog louse: *Lingogathhus setosus*).
- Buffalo louse: *Haematopinus tuberculatus.*

- Eggs are usually stuck on to the host's hair and hatch when the temperature is sufficiently high
- Head louse female can lay up to 10 tubular eggs a day. The eggs are pearly white, 0.8 x 0.3 mm. The egg hatches in 8 or 9 days, and will reach adulthood in less than 3 weeks. A female can lay up to 60 eggs, and it has been found that more than 80% of these go on to hatch. Head lice will die of starvation if they do not drink a blood meal for 55 hours. Transmission is though direct contact or sharing combs or brushes.
- Pediculosis is an infestation of lice anywhere on the human body. It is usually characterized by skin irritation, allergic reactions, and a general feeling of malaise. In addition, the human body louse is responsible for the spread of relapsing fever (*Borellia recurrentis*), epidemic Typhus (*Rickettsia prowazeki*), and trench fever (*Rickettsia Quintana*).
- Louse borne disease is particularly common in war time when soldiers were forced to live in crowded and unsanitary conditions. Trench fever was especially widespread during World War I, and was probably a major factor in the final collapse of the Russian army.
- **A 'nit' is the egg stage of a louse.** Nits are always glued to the hair of the host animal. The term "nit picking" refers to the type of detailed or meticulous efforts that would be needed to find and remove the nits from an animal's fur.

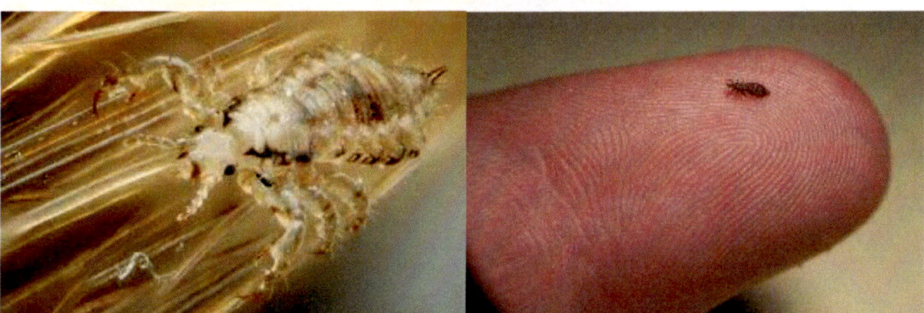

Pediculus humanus, the human louse

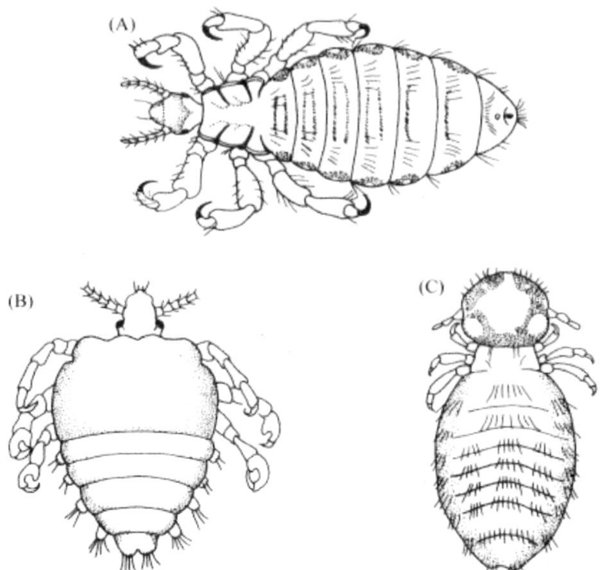

(A) *Pediculus humanus* (B) *Phthirus pubis* (C) *Trichodectes canis* of dogs.

Hemiptera (Hemi-half; ptera-wings)

The **Hemiptera**, also called "true bugs" or simply "bugs", include - true bugs; stink bugs; shield bugs; squash bugs; milkweed bugs; ambush bugs; cinch bugs; water boatmen; water striders; cicadas; aphids; scale insects; leafhoppers;

- It is the largest of the non-endopterygote orders, with more than 90,000 species in about 140 families.
- Size range from 1 mm to 11 cm in length.
- Head porrect (Heteroptera) or deflexed (Homoptera).
- Mouthparts: Piercing and sucking type. the mandibles and maxillae modified as needle-like stylets, lying in a beak-like, grooved labium, collectively forming a rostrum or proboscis within which the stylet bundle contains two canals, one delivering saliva, and the other uptaking fluid. Palps are lacking.
- In male coccids-mouthparts atrophied.
- Labium (Beak) arise from the anterior (Heteroptera) or posterior (Homoptera) part of the head.

Heteroptera

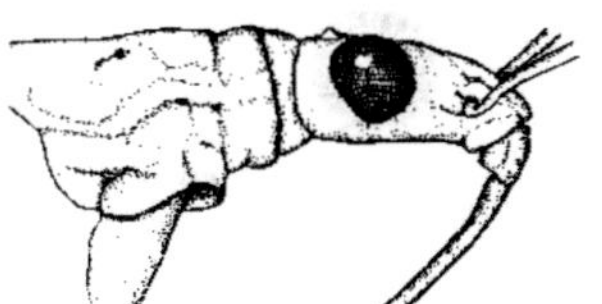

Auchenorrhyncha

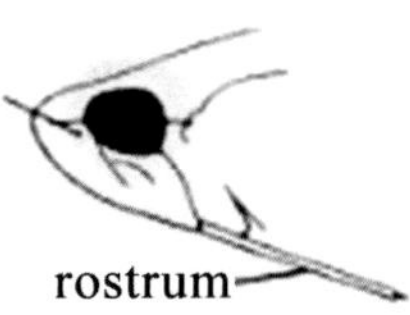

Sternorrhyncha

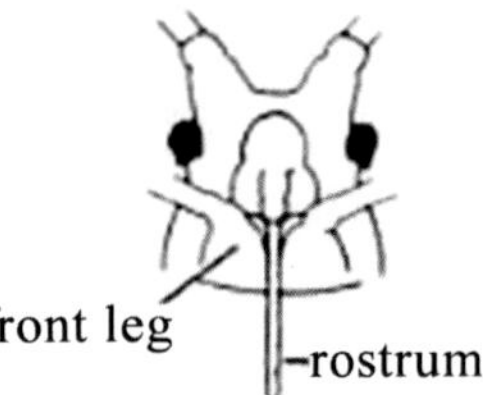

- Antenna: 4-5/10-25 segmented.
- Compound eyes are well-developed. 2 ocelli may be present/absent
- The insects in the order hemiptera have a particular structure of the front wings from which the order gets its name
    - Basal portion of the front wing is thickened and leathery
    - Apical portion is membranous (this type of wing is called hemelytron, or hemelytran if single)
    - Hind wings are completely membranous and shorter than the front wings
    - Wings at rest are held over the abdomen with membranous tips overlapping
- Wings: Hemielytra (Heteroptera), similar in Homoptera; presence of distinct scutellum between the base of wings.
- The hemelytron is composed mainly of regions (corium, clavus and membrane). In some insects, a narrow strip of corium along the costal margin is set off from the reminder of corium by a suture and is called embolium. In a few hemiptera, a cuneus is set off by a suture from the apical part of corium.
- Legs - First pair of legs raptorial in predacious forms. 2-3 tarsal segments with a pair of claws, supported by an empodium but without arolium. Pulvilli are present one under each claw.
- Cerci absent.
- Ovipositor present. **Absent in aphids and coccids.**
- Abdomen 11 segments, first 2 segments are modified for sound production (especially in cicadidae)
- Anal cerci Absent
- Ovipositor with 3 pairs of valves, absent in aphidoidea and coccoidea

- Spiracles 10 pairs present in Heteroptera
- Cicadidae – 1st and 2nd abdominal segments modified for sound production.
- Filter Chamber-Homopterans.
- Nymphs have external wingpads (sometimes absent) and unsegmented tarsi.
- Nymphs always lack wings.
- Aphids:Parthenogenetic.
- Wax glands: Aphids, coccids, Aleyrodids.
- Scent glands (*odoriferous glands*) are present, *opens near hind coxae*, give off characteristic odour in Heteroptera
- Pupal Stage: Males of coccids and both sexes of Aleyrodids (White flies).
- Hemiptera and Thysanoptera are sister groups within Paraneoptera. Hemiptera used to be divided into two groups, Heteroptera (true bugs) and "Homoptera" (cicadas, leafhoppers, planthoppers, spittle bugs, aphids, psylloids, scale insects, and whiteflies), treated variously as suborders or orders.
- **Heteroptera** (true bugs, including assassin bugs, backswimmers, lacebugs, stink bugs, waterstriders, and others)
- All **homopterans** are terrestrial plant feeders and many share a common biology of producing honeydew and being ant-attended. The wings are held roof-like over the abdomen, forewings in the form of a tegmina of uniform texture, and with the rostrum arising ventrally close the anterior of the thorax.

Economic Importance

Plant feeding bugs cause localized injury to plant tissues, they may weaken plants by removing sap, and they, may also transmit plant pathogens. Predatory species of Heteroptera are generally regarded as beneficial insects, but those that feed on blood may transmit human diseases. **Chagas disease**, for example, is transmitted to humans by conenose bugs (Genus Triatoma, family Reduviidae). Although bed bugs (family Cimicidae) can inflict annoying bites, there is evidence that they regularly transmit any human or animal pathogens. Homoptera are among the most abundant herbivores found in terrestrial habitats. Many species are pests of cultivated plants. Aphid and leaf hoppers are important carriers of plant diseases.

Sub orders

Attributes	Heteroptera	Homoptera
Size	Comparatively large	Comparatively small
Head	Porrect	Deflexed
Antenna	Slender	Slender or bristle like
Pronotum	*Pronotum Large,* while meso- and meta-notum are relatively small	*Pro-notum Small*, collar-like, large meso-notum, and small meta-notum
Rostral base	Not contacting fore coxa, long	Contacting pre coxa, shorter
Beak	Beak arises from the anterior part of the head.	Beak arises from the posterior part of the head.
Gula	Sclerotised	Absent
Tarsi	Usually 3- segmented	1-3 segmented
Wings	Not alike (Forewings -Hemi elytra); hind wings membranous	Alike (both fore and hind wings are membranous)
Wings position at rest	Overlapping(flat) the abdomen forming an "X"	Slopping over the abdomen in tent like manner
Scutellum	Well developed	Not well- developed
Thorax	Pronotum uniform usually greatly enlarged	Small and collar like
Honey dew secretion	Uncommon	Common
Disease vectors	Do not transmit diseases	Most insects act as vector
Malphigian tubules	Generally four in number. Only two in *Lethocerus*.	Variable in number.
Glands	Repugnatorial or Odoriferous or scent glands present	Wax glands usually present
Habitat	Both terrestrial and aquatic	Terrestrial
Habit	Herbivorous, Predaceous or blood sucking	Herbivorous
Life history	Parthenogenesis is present in some insects	In most species, the life history is very complex involving sexual and parthenogenetic generations, winged and wingless individuals
Members	True bugs	Aphids, Jassids, Hoppers, Whiteflies, Mealybugs , Scales, Psyllides, Cicadas
Divisions	Division: **Cryptocerata**: short antenna concealed in grooves on ventral side of the head	Series: **Auchenorryhncha**-Origion of rostrum is more cephalic in nature.
	Division: **Gymnocerata:** Long antenna not concealed in grooves on ventral side of the head (water bugs/Short horn bugs)	Series: **Sternoyyhyncha**-Origin or rostrum is more thoracic in nature.

Sub Order Heteroptera (True Bugs)

- Small to large mostlty terrestrial, some are acquatic
- Antennae fairly long, 4 or 5 segmented compound eyes well develpped. Ocelli
- when present are two in number
- Mothparts piercing and sucking type and with slender segmented beak (modified labium) that arises from the front part of the head and usually extends back along the ventral side, some times as far as the bases of hind coxae. The beak and rostrum serves as a sheath for the four piercing stylets, two outer mandibular and two inner maxillary stylets. The inner maxillary stylets fit very close together with central ridge in the groove and form dorsal food channel and ventral salivary channel. There are no palpi.
- Pronotum large, the mesonotum exhibits five fold division, among which
- scutellum is very prominent
- Winged and wingless. When winged, the fore wings are basally thickened and membranous apically and are known as Hemelytra. The hemelytron is composed mainly of regions (corium, clavus and membrane). In some insects, a narrow strip of corium along the costal margin is set off from the reminder of corium by a suture and is called embolium. In a few hemiptera, a cuneus is set off by a suture from the apical part of corium. Hind wings are entirely membranous and are slightly shorter than forewings.
- At rest the wings are held flat on the body. Alary polymorphism is seen.
- Odoriferous glands or repugnatorial glands or scent glands or stink glands are present which open near hind coxae on the sides by ventral pores giving out unplesant odour
- Ovipositor small with two pairs of valves or well developed for inserting their eggs in plant tissues.
- Anal cerci absent
- Metamorphosis simple

Family	Features
Cimicidae (Bed bugs)	• Blood sucking ectoparasites on man and animals. • Body is dorsoventrally flattened, dull reddish brown in colour. • Stink glands are located in the dorsal surface of first three abdominal segments. • Male bed bugs piece the integument of the female and inject the sperm into the haemocoel during copulation (**Heamocoelic** or **traumatic** insemination). • Bed bug-*Cimex lectularius*, *C. hemipterus* transmit Kala azar. It can produce pruritic, urticarial, vesicular, and even bullous lesions. • *Cimex lectularius* is found in subtropical and temperate zones, and *C. hemipterus*, found in tropical/equatorial areas.
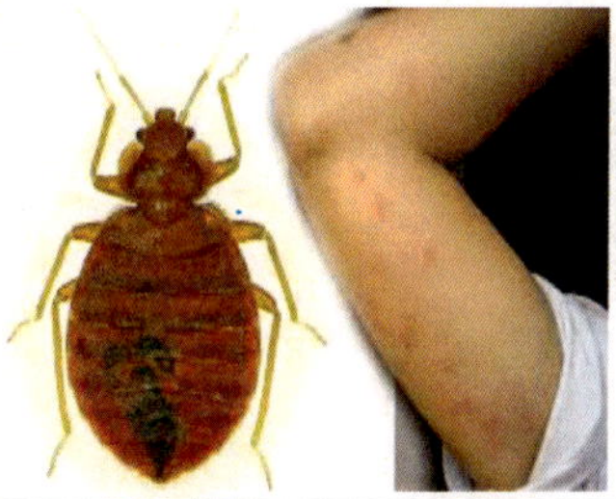	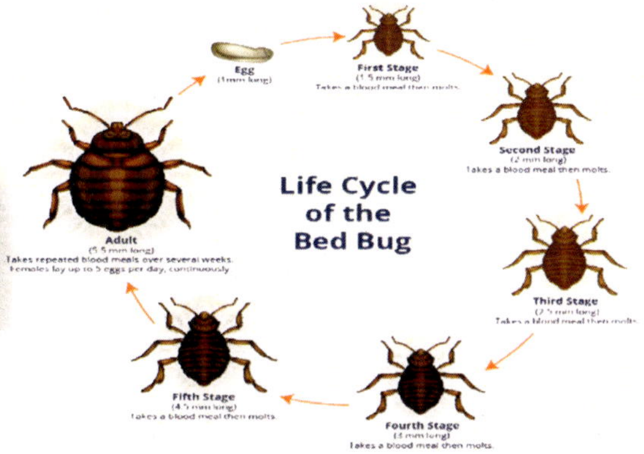
Lygaeidae (Chinch bug/milk weed bug)	• Red, white or black spots or bands. • Small bugs, hard bodied • Antennae 4 segmented inserted down on the sides of the head, apical segment is larger • Compound eyes and ocelli are well developed • 4 to 5 unbranched simple veins in the membrane of hemelytra. Cuneus is lacking, clavus is elongate • Metathoracic gland openings are present. • In some, the front femora moderately swollen with 2 rows of teeth beneath • Coxa rotator, tarsi 3 segmented, pulvulli present • Eg. Dusky cotton bug – *Oxycarenus hyalinipennis* • Groundnut pod bug –*Aphanus soridiidus* • Dusky cotton bug/cotton strainers : *Oxycarenus laetus*.

Oxycarenus hyalipennis Oxycarenus latus Aphanus sordidus

Pyrrhocoreidae (Red bugs/strainers)	• Known as cotton strainers as they pierce the bolls and make them vulnerable for contamination by fungus *Nematospora gossypii* which strains the lint. • They exhibit red and black colourations • Ocelli absent • More branched veins and ceels present in hemelytra • Coxa is rotatory and tarsi 3 segmented with pulvilli • Red cotton bug: *Dysdercus cingulatus*.

Dysdercus cingulatus

Coreidae (Leaf footed bug)	• Antennae 4 segmented situated well upon the sides of the head above a line drawn from the eyes to the base of the beak. • Ocelli present. • Head narrower, shorter than pronoutm, scutellum smaller. • In hemelytra, richly branched veins are present. • Legs are flattened leaf like (either or both the hind femora and tibiae may have conspicuous enlargements or leaf like dilations). • Tarsi – 3 segmented pulvilli present. • Metathoraccic gland openings present • Odoriferous glands on the fifth abdominal segment produce nauseous odour (but-2-enal). • Paddy Gundhi bug: *Leptocorisa acuta*. • Pod bugs :*Clavigralla gibbosa* on pulses
   *Leptocorisa acuta Clavigralla gibbosa*	
Pentatomidae (stink bugs or shield bugs)	• Medium to large insects, brightly coloured • Head with lateral margins concealing bases of antennae • Antenna is 5-segmented. • Ocelli almost always present • The pronotum broad and shield shaped. Scutellum is prominent and shield like. • In hemelytron, corium large extending to anal margin. Membrane with many longitudinal veins, arising from a vein which is nearly paralleled to the apical margin of corium. • Four pairs of odoriferous glands are present on dorsum of abdomen of the nymphs • 2-3 segmented, claws with pulvilli • The eggs are usually barrel shaped with spines on the upper end. • Adults and nymphs produce a disagreeable odour from stink glands located in metathorax and abdomen respectively. • Some are phytophagous and some are predaceous. • Predatory bugs: *Andrallus spinidens* of tobacco caterpillar. • Phytophagous mustard painted bug: *hilaris (cruciferarum)* • Green (stink) bug – *Nezara viridula*, • Red pumpkin bug -*Aspongopus janus*

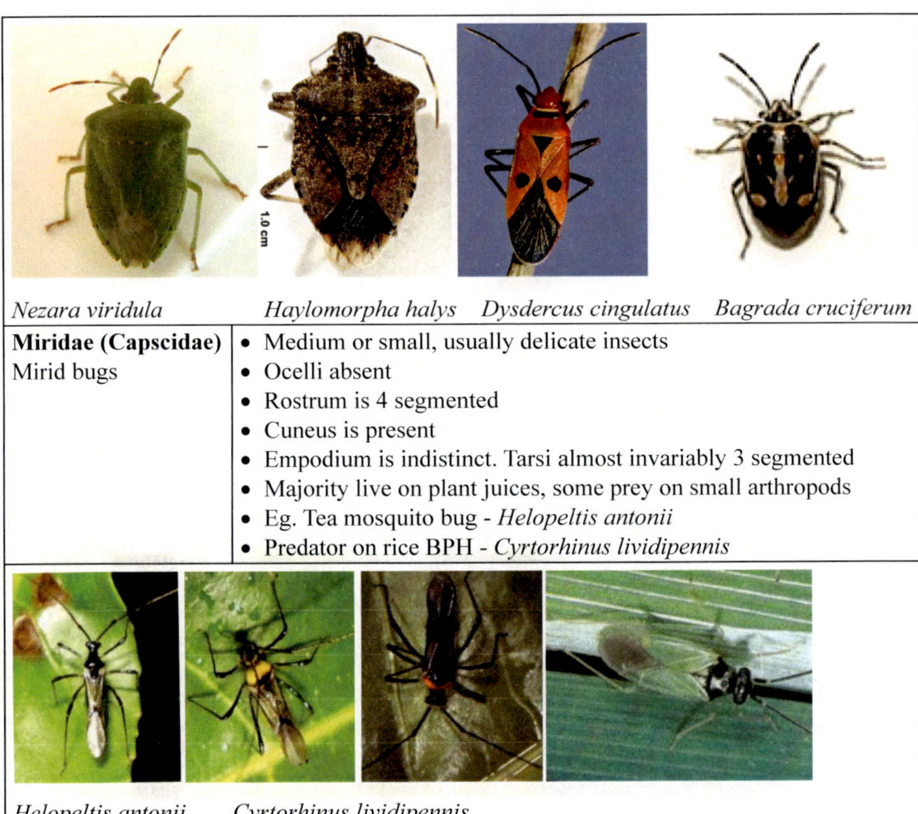

Nezara viridula *Haylomorpha halys* *Dysdercus cingulatus* *Bagrada cruciferum*

Miridae (Capscidae) Mirid bugs	• Medium or small, usually delicate insects • Ocelli absent • Rostrum is 4 segmented • Cuneus is present • Empodium is indistinct. Tarsi almost invariably 3 segmented • Majority live on plant juices, some prey on small arthropods • Eg. Tea mosquito bug - *Helopeltis antonii* • Predator on rice BPH - *Cyrtorhinus lividipennis*

Helopeltis antonii *Cyrtorhinus lividipennis*

Sub order: Homoptera

1 These are minute to small insects and are distributed widely

2 Head is deflexed and not generally constricted behind to form a neck.

3 Compound eyes well developed, ocelli absent in apterous form but 2 to 3 in winged forms

4 Antennae well developed and usually 3 to 10 segmented

5 Mouth parts piercing and sucking type, stylets often exceedingly long, retractile, rostrum arising from the back of the head, in some cases appearing to arise between anterior coxae. In some adults like male coccids the mouth parts are vestigial or absent

6 Thoracic segments generally fused together and not distinguishable from abdomen in wingless forms. Pronotum small and collar like.

7 Winged or wingless when winged the four wings are uniform in consistency and the wings are held roof like over the body at rest. Alary polymorphism

is prevalent. In male coccids only one pair i.e forewings are present.

8 Wax glands or honey tubes usually well developd in most of the members of this order.

9 In most of the species, the life history is very complex involving sexual and parthenogenetic generations winged and wingless individuals.

10 Usually undergo simple metamorphosis. In some species, the last nymphal instar is quiescent and pupa like.

Family	**Features**
Cicadellidae/Jassidae (Leaf hoppers)-Largest family	• Small size (< 1.5 cm) have the habit of running sideways. • Vector of plant diseases. • *Cause plant injury by injecting toxins*, causing hopper burns • Slender, usually tapering, posteriorly, wedge shaped insects usually rest in a position ready for jumping. When disturbed they leap often several feet. • Antennae minute, bristle like, 3 segmented • Forewings are somewhat thickened and often brightly coloured • Anal veins 1A and 2A do not unite to form 'Y' shaped vein One or two rows of small spines are present on hind tibia which is most important feature • Both young ones and adults have the characteristic habit of running sideward or diagonally • Ovipositor well developed and adopted for lacerating plant tissues for egg laying. • Many excrete honeydew through anus. • Symptoms of attack-**Hopper burn** and **honey dew secretion (Sooty mould)** **Examples** • Rice green leaf hopper (GLH): *Nephotettix viriscens* vector of Tungro,dwarf and yellow dwarf virus. • Sesame cicadellid: *Orosius alvicinctus* Vector of sesamum phyllody virus. • Brinjal cicadellid: *Cestius phycitus* Vector of little virus disease. • Mango hopper: *Amritodus atkinsoni, Idioscopus niveosparsus, I. clypealis* • Cotton/okra Leaf hopper: *Amrasca bigutulla bigutulla*

Family	Features
A. atkinsoni *Amrasca sp.*	*Green leaf hopper* *White rice Leaf hopper*
Delphacidae (Plant hopper)	• Largest family among the plant hoppers • Small insects with reduced wings • Presence of large flattened spur (Calcar) at the apex of hind tibia. • Costal cell is absent in the winged forms • Alary polymorphism is present within the same species (Apterous, Brachypterous and Winged) • Large mobile apical spur on hind tibia • Anal vein 1A and 2A unite to form Y shaped vein **Examples** • Rice white backed plant hopper (WBPH):*Sogatella furcifera* vector of rice yellow virus. • Rice brown plant hopper (BPH): *Nilaparvata lugens;* Vector of grassy stunt virus. • Corn lantern fly : *Peregrinus maidis* vector of stripe virus disease of jowar, maize.
Nilaparvata lugens	*Pergrinus maidis* Hopper burn
Lophopidae	• Sugarcane leaf hoper (*Pyrilla perpusilla*)

Family	Features
Diaspididae (Hard or armoured scales)	• The body of the adult female is sac-like, the head, thorax and abdomen not being clearly separated. • The body bears distinct dorsal gland ducts (called macroduct), which may be one- or two-barred, as well as smaller, ventral ducts (the microduct). • The head carries the vestigial antennae and the very long, needle-like mouthparts. • The thorax bears the eye spots and (in very rare cases) the atrophied legs, as well as two pairs of spiracles. • The posterior abdominal segments are fused into a pygidium, whose margins carry lobes and plates, of great systematic value. • The vagina is usually surrounded by genital pores, which are lacking in viviparous species (e.g. *Aonidiella aurantii*). • The body of females and males is covered by a variously-colored detachable shield, which usually differs between the sexes, being larger in the female. The dorsal part of the exuvium of the preceding stages is embedded within the shield. **In contrast to other Coccoidea, the shield of the Diaspididae is removable** (except at molt and consists of waxes and a polyphenolic polymer. • Females have two immature instars whereas males have four, including the prepupa and pupal stages, which remain within the exuvium (velum) secreted by the second instar. • The first instar larvae (usually called crawlers} beneath its body.

Aonidiella auranti

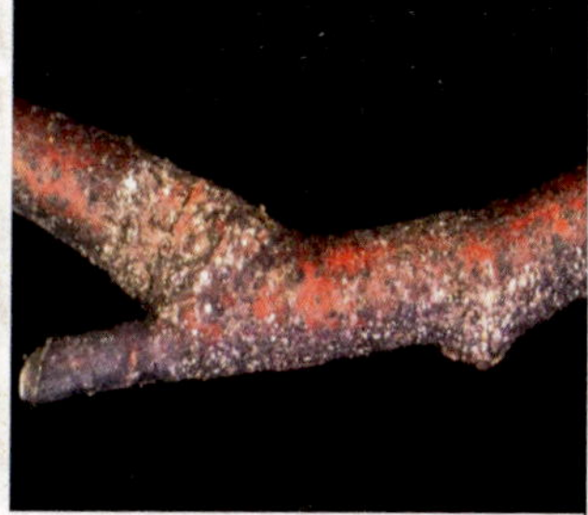

Quadraspidiotus perniciosus

Family	Features
Aleyrodidae (white flies)	• Minute (1-3mm), white colour flies mostly found on under surface of leaves. • Body and wings of adults are covered with a white powdery wax resembling tiny moths with opaque body. • Compound eyes well developed and ocelli two in number • Antenna 7-segmented. • Rostrum 3- segmented. • Tarsi 2- segmented with paired claws and with empodium. • Oviparous. • 4th instar is stationary called as pupa. • Vasiform orifice present on the dorsal surface of the last abdominal segment within which is situated on operculum is of taxonomic significance. • Metamorphosis is complex. The 1st instar young ones are active (crawlers) but subsequent immature stages are sessile and look like scales. The scale like covering is a waxy secretion of the insect. The wings develop internally during metamorphosis and the early instars are called larvae. The next to the last instar is quiescent stage called pupa. The wings are given out at the moult of last larval instar. • The eggs are very characteristic being provided with pedicel which sometimes exceeds the length of the egg. • Honey dew is excreted in large quantities particularly by larvae through anus. • Cause Mosaic diseases **Examples** • Sugarcane white fly (*Aleurolobus barodensis)* • Cotton white fly (*Bemesia tabac*i): vector of leaf curl virus disease of tobacco, cotton. • Citrus Whitefly : *Dialeurodes disperses* • Spiralling Whitefly : *Aleurodicus citri disperses* • Citrus Blackfly : *Aleurocanthus woglumi* • Rose Blackfly : *Aleurocanthus spiniferus* • Castor Whitefly : *Trialeurodes ricini*

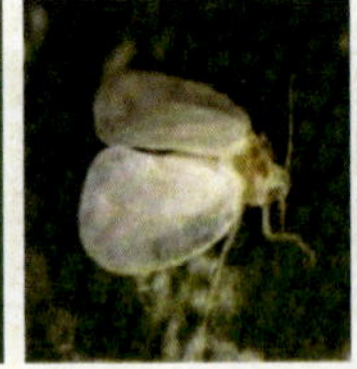

Whiteflies **Spiralling whitefly**

Family	Features
Aphididae (Aphids, green flies or plant lice)	• Second largest family in the sub-order Homoptera. • Many of these insects are pests of cultivated plants. Vectors of viral plant diseases. • Sexual forms viviparous; Parthenogenetic females Viviparous. • Pear shaped, • 3-6-segmented antenna, • several jointed rostrum, • Tarsi 2-segemented with paired claws. • Pear shaped, small and soft bodied insects • Rostrum usually long and well developed • A pair of cornicles (Honey tubes) on the 5 or 6th abdominal segment. (cornicles produce wax substance to protect from other insects) • Alary polymorphism (winged / wingless forms, normally wingless forms are predominant. In winged forms, hind wings are much smaller with fewer veins) • At rest, the wings are generally held vertically above the body. • Excrete honeydew through anus (honey dew consists of excess sap, excess sugars and waste materials), to which ants are attracted. The transportation of aphids through ants is called Phoresy. • Aphids generally exhibit polymorphism in their development in different generations of the same species. • The reproduction can be through parthenogenesis, oviparity, viviparity. • Occurrence of generations the sexes are unequally developed, males often being rare. **Life Cycle.** • Eggs • Fundatrices: apterous, viviparous, parthenogenetic, legs, antenna not developed. • Fundatrigeniae: Parthenogenetic,apterous,viviparous. • Migrantes: winged, viviparous • Aliencolae: Parthenogenetic, viviparous, winged • Sexuparae:Parthenogenetic,viviparous,winged. • Sexuales: Sexual male and females, oviparous. • Maize aphid (*Rhophalosiphum maidis*): Vector of grassy shoot and mosaic virus in sugarcane and chirke virus in cardamom. • Banana aphid (*Pentalonia nigronervosa*): Vector of bunchy top virus, cardamom mosaic virus. • Cotton aphid (*Aphis gossypii*) : vector of leaf roll and ‘ Y’ viruses in potato. • Bean aphid (*Aphis craccivora*): vector of common broad bean mosaic of beans. • Wooly apple aphid (*Eriosoma lanigerum*): Winter migration (roots) and summer migration (upper plant parts).

Family	Features

Aphis gossypi *Sitobion avenae* *Myzus persicae* *Aphis craccivora* *Eriosoma lanigerum*	
Pseudococcidae (Mealy bugs)	• Females are wingless, elongate oval with distinct segmentation. • Body covered with powdery wax or filamentous waxy secretions. • Antennae present • Legs well developed. • Instar is movable. All the insect stages are able to move because of legs • Eggs are placed in a loose cottony waxy material • Rice mealy bug (*Brevennia* (Brevennia) *rehi*): Cause disorder **Soortai** in Tamil nadu. • White tailed mealy bug -*Ferrisia virgata* • Brinjal mealy bug – *Coccidohystrix insolita* • Citrus mealy bug - *Planococcus citri* • Sugarcane mealy bug - *Saccharicoccus sacchari* • Grapevine Mealy Bug : *Maconellicoccus hirsutus* • Guava Mealy Bug / Scale : *Pulvinaria psidi* belongs to family Coccidae delete from here

Saccharicoccus sacchari *Planococcus citri* *Phenacoccus madeirensis* *M.hirsutus*	

Family	Features
Coccidae/Lecaniidae (Soft scale, Wax scale, Tortise scale)	• Female are flat elongated oval with the body smooth or covered with wax. • Legs present or absent. • Antenna up to 13-segemented but in females reduced or absent. • Males may be alate or Apterous. • The females in this group are flattened, elongate oval insects with obscure segmentation and hard smooth exoskeleton or covered with wax or tough scales. They are wingless, legs present or absent and the antennae absent or much reduced. • Males are active, 1st pair of wings well developed, 2nd pair reduced to halters. • Tarsus if present 1 – segmented with a single claw. • Metamorphosis complex. **1st instar** nymph has legs & antennae and active known as **crawlers** after 1st moult, become sessile a waxy or scale like covering is secreted. • In males, last instar preceding adult is **quiescent** and called **pupa**. Females have one less instar than males. • Oviparous, ovoviviparous • Excrete honey dew like aphids through anus • Anus covered by 2 dorsal plates **Examples** • Sugarcane Scales: *Melanapsis glomerate* belongs to family Diaspididae • Cottony Cushiony Scale/Citrus Scale : *Icerya purchasi* • Mango Scale : *Aspidiotus destructor* belongs to family Diaspididae • Guava Scale / Mealy Bug : *Pulvinaria psidi* • Black Scale on Crotons : *Saissetia nigora Parasaissetia nigra* • Coffee Scale : *Saissetia coffeae* • Citrus Scales : *Aonidiella citrina* belongs to family Diaspididae

Female	**Male**
Females are very common	Males are rare
The females in this family are degenerated, flattened, elongate, oval insects with obscure segments. Hard smooth exoskeleton or covered with wax or tough scales	Males are active
Wingless	Front pair of wings highly developed, Hind pair reduced to halteres
Legs are present or absent	Tarsus if present, 1 segmented with single claw
Antennae absent or much reduced	Antenna long and hairy
Mouthparts functional, well developed	Mouth parts vestigial or absent
Adults feed on plant sap	Adults do not feed and short lived
	Caudal filament present at tip of the abdomen

Family	Features
Cottony cushion scale	*Pulvinaria* sp. *Aspidiotus*
Saissetia coffeae	

Thysanoptera

(Thysano-fringe, ptera-wings; **Thrips**)

- The name Thysanoptera derived from the Greek "*thysanos*" meaning fringe and "*ptera*" meaning wings, refers to the slender wings that bear a dense fringe of long hairs.
- Thrips are generally small (under 3 mm), yellow brown or black insects having habit of hiding in flowers and flower buds
- Front and hind wings slender, rod like, with a dense fringe of long hairs. Many species are secondarily wingless.
- Flight-capable thrips have two similar, strap-like, pairs of wings with a ciliated fringe, from which the order derives its name. Wings when fully developed are long and narrow with highly reduced venation (with 1-2 veins). The wings are fringed with long hairs on the margins.
- Hemimetabolous.
- Compound eyes conspicuous with 3 ocelli in winged forms
- Antennae short, 6-10 segments moniliform type with sensorial (cone like structure) on 3rd or 4th segment
- Tarsi 1-2 segmented with eversible adhesive bladders apically. Thrips are able to walk on glossy smooth surfaces because they have an eversible adhesive pad located on the tip of each foot, between the claws.
- Cerci absent.

- Thrips are the only insects that have asymmetrical mouth parts. Only left mandible is present and functional. Right mandible is absent.
- Thrips are the only members of the 'Exopterygote' to have developed a true pupal stage. They are described as holometabolous (having a complete metamorphosis even though the nymphs look like small wingless adults) insects with 2 or 3 inactive pupa like instars.
- Intermediate between simple and complex, in which 1st and 2nd instar larvae are active, resemble adults (called nymphs) where as 3rd and 4th instar consists of inactive pre-pupal and pupal like stages.
- Parthenogenesis is very common, and in many species males are rarely seen.
- Many thrips are destructive pests of plants especially grain crops, fruits and vegetables, and ornamentals. Feeding activities result in plant deformities, scarring, loss of yield and in some cases, transmission of plant pathogens. Predatory thrips are beneficial species that may control mites and other small insects.
- The family is divided into two Sub orders (Terebrantia and Tubulifera). The Sub order can be distinguished by the shape of ovipositor, in Terebrantia it is sharp and long but it is tubular in Tublifera.

Sub Order : Terebrantia	**Tubulifera**
Ovipositor present.	Ovipositor absent.
Last abdominal segment: Short and pointed.	Last abdominal segment: Long and tubular.
Wings with microtrichia	No venation.
Family	Family
Thripidae: Rice Thrips: *Thrips oryzae* Cotton Thrips: *Thrips tabaci*	Phaleothripidae: Black hunter, *Leptothrips mali.*

Family : Thripidae

Chilli thrips : *Scirtothrips dorsalis*

Onion thrips : *Thrips tabaci*

Grapevine thrips : *Rhapiphorothrips cruentatus*

- Thripidae is the most speciose family of thrips, with over 290 genera representing just over two thousand species.
- Largest, most important and most injurious family in thysanoptera.
- Antennae 6-9 segmented, 3rd and 4th antennal segments are conical with sense cones or sensorial, antennae with 1-2 segmented apical style, 4th segment is usually enlarged

- Winged or wingless forms. If winged, wings are narrow, pointed at the tip and fringed with hairs on the margins
- They can be distinguished from other thrips by a saw-like ovipositor curving downwards
- 1st and 2nd segments of tarsi consists of claw like appendage

Onion thrips ***Thrips tabaci***

Neuroptera (Neuro-nerve; ptera-wings)

Lacewings/Antlions/Dobsonflies/Alderflies/Snakeflies / Planipennia

The name Neuroptera is derived from the Greek word "neuron" meaning Sinew and "ptera" meaning wings. The modern English translation "nerve-wings" is appropriate because it alludes to the extensive branching found in the wing veins of most Neuroptera.

- Small to rather large soft bodied insects with usually elongated filiform multisegmented antenna.
- Mandibulate mouthparts.
- Front and hind wing membranous, similar in size.
- Relatively large compound eyes
- Often long, filiform antennae, which may be clubbed in some species
- Extensive branching of venation in all wings; cross veins abundant especially along leading edge (Costal margin). At rest, the wings are folded flat over the abdomen or held tent like over the body. Most species are rather weak fliers.
- No cerci.
- **Larva** carnivorous, modified grub (campodeiform) well developed head well developed with well developed head and ocelli, antennae and chewing or pinching (Pincers) mouthparts.
- the mandibles and maxillae are modified to form elongate, unsegmented stylets adapted for piercing and sucking. Three pairs of thoracic legs; tarsi 1-segmented; Claws paired.

- Aquatic forms have thread like gills on most abdominal segments.
- First-instar larvae obtain oxygen from the water through their cuticle, but second- and third-instar larvae have two- or three-segmented gills on abdominal segments 1–7. These abdominal gills can be vibrated rapidly to increase water current across gill surfaces. Third-instar larvae shed their gills prior to emergence from the aquatic environment.
- Pupae exarate, decticous.
- The Megaloptera are always aquatic as immatures. They live under stones or submerged vegetation and feed on a variety of small aquatic organisms. Large species,often called **hellgrammites**, may require several years of growth to reach maturity.
- Lace wings larvae are beneficial as predators of agricultural pests (Aphids,Whiteflies and scale insects). *Chrysoperla carnea* is important as biocontrol agent.
- A lacewing`s egg sits atop a slender stock secreted by the female`s reproductive system.
- As larvae, lacewings and antlions donot have a complete digestive system. Waste material accumulate in the mid gut throughout larval stage is expelled only after a connection is made with the anus near the end of pupal stage. The accumulated fecal material is a meconium.
- Cocoon`s silk is produced by Malpighian tubules (excretory organs) and spun from the anus. Only one other order, the Coleoptera, makes silk in the same manner as Neuroptera.
- Chrysopidae (Green lacewings)- *Chrysoperla carnea*: Aphid predators.

Lacewings may be confused with dobsonflies or alderflies (Megaloptera) but can be distinguished from these insects by the presence of forked veins.

They may also be confused with stoneflies (Plecoptera) and dragonflies (Odonata) but lack the two thin abdominal cerci that stoneflies possess and usually have longer antennae and softer bodies than dragonflies.

Adult

Eggs (stalked)

Larva (Predator)

Pupa

16

Lepidoptera, Diptera, Hymenoptera and Coleoptera

The name Lepidoptera derived from the Greek words "lepido" meaning Scale and "ptera" meaning wings refers to the flattened hairs (Scales) that cover the body and wings of most adults. Lepidoptera (Moths and Butterflies) is the second largest order (1.10 lakh sp.) in the class insecta, characterized by the body, wings and appendages being generally covered densely with pigmented scales which provide the colour pattern of the characteristic of the species and play an important role in courtship and intraspecific recognition.

Adults

- Head: Hypognathus and relatively small.
- Compound eyes are large and set well apart. Ocelli two in number in moths or reduced to **Chaetosoma** in butterfly.
- Mouthparts form a coiled tube (Proboscis) beneath the head.
- Antennal type:
  - Butterflies: Clubbed, clavate, capitate.
  - Moths: Pectinate (Uni/Bi).
- Adults have siphoning type mouthparts except in micropterygidae (Mandibulate mouthparts). Front wings large, triangular, hindwings large, fan shaped.
- Body and wings covered with small, overlapping scale.
- Thoracic segments are fused and metathorax is the largest and most prominent.
- Wings covered with scales. Androconia (Plumules) on upper surface of wings in some males serves as outlet for odouriferous glands. are fringed distally with each tip finely divided.
- Wing coupling apparatus variable (generally Frenulum and retinaculum or Amplexiform type)

- Females of certain geometridae and psychidae are apterous.
- Tarsi 5-segmented ending in paired claws. Forelegs reduced (Nymphalidae) or atrophied (female bag worm of Psychidae).
- Abdomen 10-segmented.Tympanum on either side of abdomen in some.
- Egg laid singly or in groups.
- Cerci absent.
- Ovipositor present.
- Adults feed on nectar flowers **except fruit sucking moths** which suck the fruit juice and cause damage to them.

Larva

- Eruciform (caterpillar-like), 3 thoracic and 10 abdominal segments.
- Head capsule well-developed with chewing mouthparts.
- 6-Ocelli.
- Antenna 3-segmented.
- Spiracles (1+9 arrangement), peripneustic.
- Mouthparts biting and chewing type (mandibulate), modified for lacerating the tissues (leaf mining larvae) or may be wanting in *Phyllocnistis citrella* (citrus leaf leafminer).
- Abdomen with up to 5 pairs of prolegs except Micropterygidae having 8 pairs of prolegs. The 1st four pairs are the abdominal feets and remaining pairs are known as claspers.
- Larvae of certain leaf miners (*Phyllocnistis citrella*) are apodus.
- Repugnatorial glands known as Osmeteria on 1st thoracic segment in Papilionidae.
- Ventral defensive glands: Noctuidae.
- Irritating/ poisonous hairs occur in Cochlidiidae, Arctiidae.
- Protective case in Bagworms (Psychidae), mimicry (Geometridae).

Pupa

- Decticous exarate in primitive families but adecticous obtect in most Lepidopterans.

- Butterflies pupa- **Chrysalis** Naked and protectively coloured supported by silken girdles.
- Moths pupa formation in cocoon.

Economic importance

- The Butterflies and Moths are by far the most popular group of insects in both the mind of the general public and with entomologists, there are more books on Lepidoptera, and more people collecting and working on Lepidoptera than any other insect order, everybody loves Butterflies.
- The Lepidoptera are one of the five great orders of insects and when all the counting has been finished will probably be fighting for 3rd place with the Diptera behind the Hymenoptera and the Coleoptera, but ahead of the Hemiptera (especially in those taxonomic organisations that split the Hemiptera into 2 orders).
- There are about 1,50,000 named species most (more than 85 percent) of which are Moths.
- The insects belong to this order, has great economic importance. Most of its members are crop pests. Some are productive insects and are useful (silkworm). Some are pollinators and some are predators.
- Small to large sized insects with overlapping scales and hairs on the body, wings and other appendages, giving beautiful colors to the insects.

Caterpillar	**Semiloopers**	**Loopers**
5 pairs of Prolegs on 3 to 6 and 10 abdominal segment.	3 pairs of prolegs on 4, 5 and 10th abdominal segment.	2 pairs of Prolegs on 6 and 10th abdominal segments.
Pieridae, most moths and butterflies.	Noctuids (Noctuidae)	Geometridae.

Difference

Characte	Butterflies	Moths	Skippers
Members	Butterflies	Moths	• Belongs to family Hesperiidae. • Widely separated at their base and each antenna is apically prolonged beyond the club into a hook or a small recurved point. • Erratic darting flight. • Larva is very peculiar with a relatively big head attached to a neck like or constricted "Collar" • Larva web the leaves or form galleries with silken thread and remain inside them. • The pupa is enclosed in the leaf in a slight cocoon or may be naked and attached by the anal end and also by a median band of silk • Rice skipper: *Pelopidas mathias*
Antennae	Clubbed never Pectinate/clavate/ capitate	Pectinate, plumose in males	
Ocelli	Absent	Mostly Presen. Generally, 2 in number	
Mandibles	Absent	Reduced	
Frenulum	Wanting	Present	
Humeral lobe of wings	Greatly developed	Undeveloped	
Wings at rest	Folded vertically upward	Roof like transversally over abdomen	
Coupling apparatus	Amplexiform	Frenate type predominant	
Cu2 vein of forewing	Wanting	Present	
Abdomen	Comparatively small	Large	
Pupa	Naked (chrysalis), coloured supported by silk girdles	Within cocoons	
Habit	Diurnal	Nocturnal	
Colour	Brightly coloured	Pale white/dull coloured	
Eggs	Cigar shaped and cylindrical	Flat and round	
Egg laying	Singly generally	Singly/groups	
Larva	Smooth and naked	Hairy	

Sub-Orders

Characters	Zeugloptera	Monotrysia	Ditrysia
Mandibles	Functional (Mandibulate)	Non-functional(Siphoning type)	Non- functional (Siphoning type)
Maxillae	Lacinia well developed	Lacinia absent	Lacinia absent
Larvae	8- Pair of abdominal prolegs	Not more than 7 pair of abdominal prolegs	Not more than6 pair of abdominal prolegs
Wings	Primitive venation	Nearly always aculeate	Not aculeate. Veination of forewings are not alike
Examples	Micropterygidae (*Micropteryx sp.)*	Hepialidae (swift moths) single genital opening	Noctuidae and all families of agricultural importance Generally have two genital openings

Major Families

Family	Features
Gelechiidae	• Small to minute moths, usually cryptic coloured. • Labial palpi are long and **upcurved**, and the terminal segment is long and pointed. • Moths forewings are trapezoidal and narrrower than hindwings. Forewing veins R4 and R5 are stalked at the base. Hind wings usually have the outer margin, curved and R5 and M1 are stalked. • Pink bollworm : (*Pectinophora gossypiella*) • Potato tuber moth:*Phthorimaea operculella* • Angoumois Grain moth:*Sitotroga cerealella* • Groundnut Leaf miner : *Aproaerema modicella*
     *Pectinophora gossypiella Sitotroga cerelella* PTM GLM	
Oecophoridae (Cryptophasidae)	• Moths without antennal pectin • Larva concealed in shelters/covering/tunnel in wood • Coconut Black Headed caterpillar: *Opisina arenosella*

Opisina arenosella

<table>
<tr><td>Phyllocnistidae</td><td>• Adult moths are very small and whitish with silvery iridescent scales, black and tan markings and a black spot on wing tip.
• Larvae apodus and mine in to leaves
• Pupation usually takes place near or on the margins of leaves.
• Citrus leaf miner:Phyllocnistis citrella</td></tr>
</table>

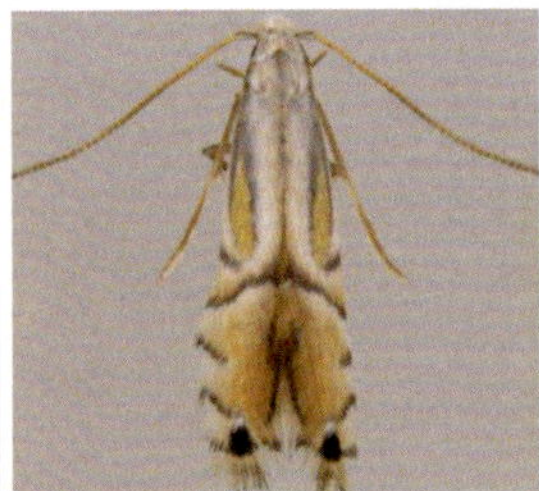

Phyllocnistis citrella

<table>
<tr><td>Plutellidae</td><td>• Adults small to medium size (7–55 mm wingspan), with head usually smooth-scaled; haustellum naked
• Labial palpi porrect (rarely upcurved); maxillary palpi 4-segmented (rarely 1–2-segmented)
• Antennae sometimes thickened at middle.
• Wings elongated, sometimes with longer fringes on hindwings, and forewings sometimes appearing apically falcate due to fringe arrangement.
• Adults mostly nocturnal or crepuscular, but some are diurnal. Larvae are leaf skeletonizers, but most remain unknown biologically.
• Diamond back moth (DBM):Plutella xylostella-a national pest of crucifers. Its wings are pale white with inner margin yellow. When the wings are folder on its body at rest, a diamond shaped median dorsal patch is seen and hence the name DBM.</td></tr>
</table>

Plutella xylostella

Tinaeidae(cloth moths)	• Wrapper tobacco larva : *Demobrotis sp.* • Webbing clothes moth : *Tineola bisselliella* • The case making clothes moths : *Tinea pellionella*
  *Tinea pellionella Tineola bisselliella*	
Metarbelidae (Barkborers)	• Larvae are wood borers .Moths are nocturnal • Guava Bark eating caterpillar/Bark borer*: Indarbela tetraonis*
   *Indarbela tetraonis Indarbela* sp.	
Galleridae (Was moths)	• Larva feed on wax of honeybee combs and wasp`s combs • Greater wax moth:*Galleria melonella* • Lesser wax moth:*Achroia grisella*

<table>
<tr><td colspan="2">  
Achroia grisella

Greater wax moth</td></tr>
<tr><td>Crambidae (Grass moths, Chilo sp.)</td><td>• Labial palps are extended.
• Forewings are narrow and elongated. At rest they are wrapped around the body.
• Eggs laid in 1-2 rows not covered with silken hairs
• Larvae remain in silken webs or galleries and bore in plant parts
• Sorghum stem borer: Chilo partellus
• Sorghum shoot borer: Chilo infuscatellus
• Legume pod borer : Maruca vitrata
• Sugarcane internode borer : Chilo sacchariphagus indicus</td></tr>
</table>

Maruca vitrata

Chilo partellus

Chilo infuscatellus

Chilo sacchariphagus indicus

Phycitidae	• Pulse pod borer:*Etiella zinckenella* • Brinjal stem borer:*Euzophera perticella* • Sapota leaf webber:*Nephopteryx eugraphella*

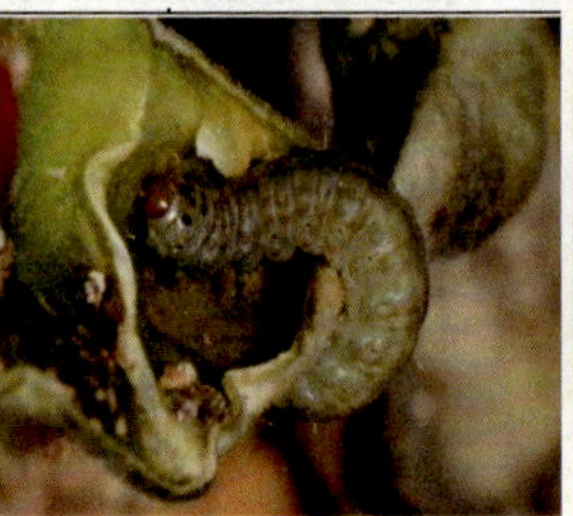

Etiella zinckinella

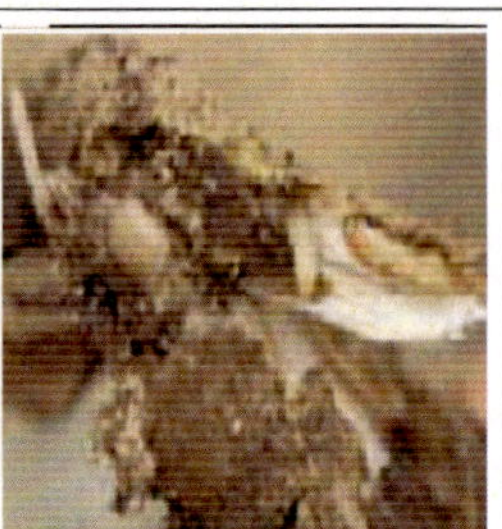

Euzophera perticella

Pyralidae (Pyraustidae) Snout moths	• These are small and delicate moths with well developed antennae and ocelli present • Vestigial proboscis • Labial palpi well developed and projected forward appearing as a snout in front of the head and hence are called snout moths. • Forewings elongate or triangular with cubitus appearing four branched. Hind wings are usually broad with SC and R usually close together, fused or closely parallel for a short distance beyond discal cell. • Generally internal feeders. • Adult females are generally provided with a tuft of anal hairs at the caudal extremity which are deposited as a covering on the egg masses • Larvae are naked, prolegs variable but always present in VIth segment. • Larva feed by boring, webbing or rolling leaves **Examples** • Rice yellow stem borer: *Scirpophaga incertulas* • Sugarcane top borer:*Scirpophaga excerptalis* • Sesame leaf webber:*Antigastra catalaunalis* • Rice leaf roller folder:*Cnaphalocrocis medinalis* • Castor shoot and capsule borer:*Dichocrocis punctiferalis* • Cabbage borer: *Hellula undalis* • Brinjal fruit and shoot borer:*Leucinodes orbonalis* • Brinjal Stem Borer : *Euzophera perticella* • Cotton leaf roller: *Sylepta derogata* • Crucifer leaf webber : *Crocidolomia binotalis* • Sapota leaf webber : *Nephopteryx eugraphylla* • Apple and nut borer of cashew :*Thylacoptila paurosema* • Mango Leaf Webber : *Orthaga exvinacea*

Chilo partellus Scirpophaga incertulas Scirpophaga innotata

Leucinodes orbonalis

Leucinodes orbonalis *Cnaphalocroscis medinalis*

Sylepta derogate *Orthaga exvinacea*

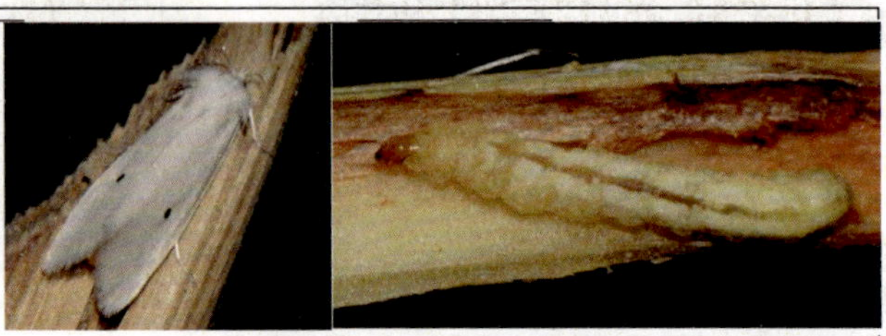

Nephopteryx eugraphella

Scirpophaga excerptalis

Hellula undalis

Crocidolomia binotalis

Dichocrocis punctiferalis

Apple and nut borer of cashew, *Thylacoptila paurosema*

Saturniidae(Silk worms)	• They are large sized moths. • Antenna is bipectinate. • Transparent eye spots are present near the centre of each wing. The spots are either circular or crescent shaped. • Larva is stout and smooth with scoli. • Cocoon is dense and firm. • Tasar silkworm:*Antheraea paphia* • Eri silkworm:*Philosamia ricini* • Muga silk worm:*Antheraea assama*

Antheraea paphia *Antheraea assamensis* *Philosamia ricini*

Bombycidae (Mulberry silkworm)	• Antenna is bipectinate. • Larvae is either with tuft of hairs or glabrous with medio dorsal horn on the eighth abdominal segment. • Pupation occurs in dense silken cocoon. • Mulberry silkworm:*Bombyx mori*

Bombyx mori

<table>
<tr><td>Nymphalidae (Brush footed butterflies)</td><td>• Forelegs reduce/non functional
• Known as 4 reduced butterflies
• Foretibia is short and covered with long hairs.
• Larva is with many processes or spines on the body.
• Rice horned caterpillar, common evening brown: Melanitis leda ismene
• Castor butterfly : Ariadne merione</td></tr>
<tr><td colspan="2">  
Melanitis leda ismene</td></tr>
<tr><td>Lycaenidae (Blues, Coppers, Hair streaks)</td><td>• Colourful tail like prolongation on hindwings
• Medium sized butterflies with upper surface of wings being metallic blue or coppery, dark brown or orange and under surface more sombre with delicate streaking or dark centred eye spots. Hind wings are provided with delicate tail like prolongations
• The sexes frequently exhibit great differences in colouration, the male is pale shining blue and the female is iridescent brown
• Each compound eye is surrounded by a rim of white scales and the antennae ringed with white
• Legs are normal except fore legs of males which may possess more or less shortened tarsi and may be wanting in one or both claws
• Larvae are characteristically vasiform with both ends tapering end with broad projecting sides concealing the legs
• Pupa is attached to the surface by its anal end and is held by a central girdle of silk, rarely it is subterranean
• Larvae are voracious feeders, some species are carnivorous
• Anar fruit borer : Deudorix (Virachola) isocrates
• Red gram blue butterfly : Lampides boeticus</td></tr>
<tr><td colspan="2">  
Virachola Isocrates</td></tr>
</table>

  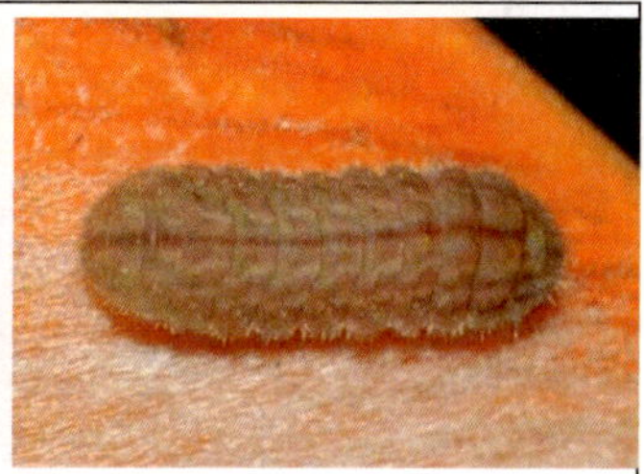

Lampides boeticus

Pieridae (whites)	• They are white or yellow or orange coloured with black markings. • Larva is green, elongate and covered with fine hairs. • Larval body segments have annulets. • Cabbage butterfly/caterpillar : *Pieris brassicae* • Indian cabbage white : *Pieris candida* • Dhaincha caterpillar: *Eurema hecabe* • Common grass yellow : *Eurema* sp.

Pieris brassicae

 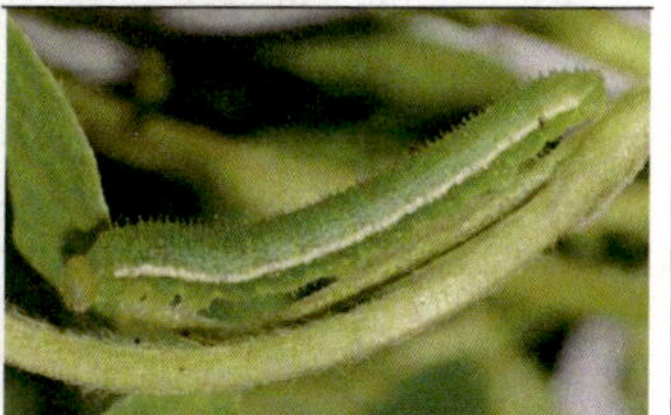

Eurema sp. (Common grass yellow)

Pieris candida

Papilionidae (Swallow tails, parnassians, Common lime butterfly, lemon butterfly, swallowtail)	• Medium to large sized butterflies. • Tail like prolongation on hind wings in most of butterflies, while parnassian adults are medium, tailless, and have translucent wings. • The cervical sclerites join beneath the neck. • Cubitus of the fore wings appears 4 branched. • The second anal vein (vein 2A) on the adult forewing extends to the wing margin and does not converge with the first anal vein (vein 1A). In all other butterfly families veins 1A and 2A fuse, and 2A does not reach the wing margin.
	• Caterpillars possess a forked, eversible organ behind the head known as an osmeterium. The osmeterium secretes a foul-smelling terpene-based defensive compound. When disturbed, a caterpillar will evert its osmeterium and rear its head back in an attempt to dissuade its antagonist. • Larva is smooth with a series of fleshy dorsal tubercles or occasionally a raised prominence on the 4th segment. • Pupa have 2 lateral cephalic projections and in hindwings being visible ventrally **Examples** • Lemon butterfly:*Papilio demoleus* • Curry leaf butter fly : *Papilio polytes*

Papilio demoleus

Papilio polytes

Osmetarium – a defensive organ

<table>
<tr><td>Geometridae (True looper, Carpets, Waves)</td><td>• Both pairs of wings are angular and thin.
• Larva is naked and elongate. It shows protective resemblance to twings or stems. Only two pairs of prolegs are present in sixth and tenth abdominal segments. It walks by drawing the posterior part of the body close to the thorax, the body forming a loop.
• It is also called inch worm, measuring worm and earth measurer.
• Dhaincha looper:Semiothisa pervolgata
• Tea looper, Biston suppressaria.</td></tr>
<tr><td colspan="2">  
Biston supressaria</td></tr>
<tr><td>Limacodidae (Slug caterpillar)</td><td>• They are medium sized moths with stoutly built body.
• Larva resembles the slug. Larva is thick, short, fleshy and stout. Larval head is small and retractile. Thorocic legs are minute. Abdominal segmentation is indistinct.
• Prolegs are absent. Poisonous urticating hairs are present on the body.
• Pupal cocoon is hemispherical with urticating hairs.
• Castor slug caterpillar : Parasa lepida.
• Green nettle slug caterpillar:Thosea aperienus also known as "leaf scorpions" in south-India</td></tr>
<tr><td colspan="2"> 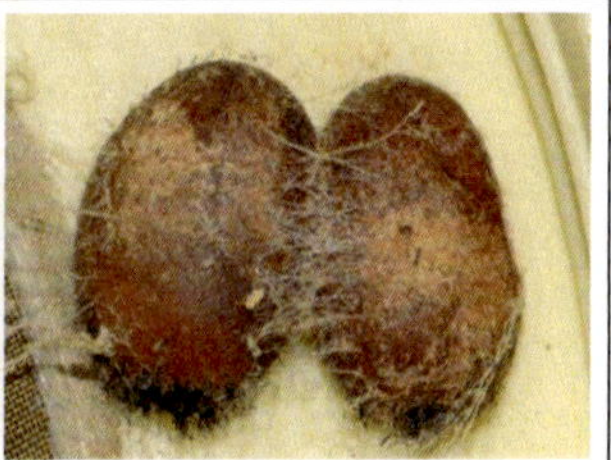
Parasa lepida</td></tr>
</table>

<table>
<tr><td>Sphingidae (Sphinx moth/ Death`s Head moth)</td><td>• Medium to large sized, heavily bodied, powerfully flying moths with spindle shaped body tapering and pointed both anteriorly and posteriorly.
• Antenna is thickened in the middle or towards the tip. Usually pointed or hooked at the tip.
• The proboscis is very long in most of the species, and attains its greatest length in this family. Usually adults produce sound by forcing air through proboscis.
• The fore wings are elongate with oblique outer margin. Hind wings small and usually lightly coloured.
• Larvae of most of the species have a conspicuous horn on the dorsal surface of 8th abdominal segment with is relatively longer in 1st instar, hence the name Horn Worms.
• Pupation occurs freely in a cell in the ground or in a very loose cocoon on the surface among leaves. In some genera the proboscis projects from the body resembling the handle of a pitcher.
Examples
• Potato sphingid : Acherontia styx
• Tobacco hornworm : Manduca sexta
• Sweet Potato Sphinx : Agrius (Herse) convolvuli</td></tr>
<tr><td colspan="2">  
Potato sphingid:Acherontia styx</td></tr>
<tr><td>Arctiidae (Tiger moths, Footman moth, flag moths and wasp moths)</td><td>• These are stout bodied, medium sized, conspicuously and brightly spotted or banded moths
• Nocturnal
• Sc and Rs in hind wings are usually fused near or beyond the middle of the discal cell.
• Larva densely clothed with hairs and some are called wooly bears and they curl into a compact mass when disturbed
• Many species are capable of producing sound
• Pupation in cocoon and the cocoons are made of silk and larval body hairs.
Examples
• Bihar/jute hairy caterpillar : Spilosoma obliqua
• Red hairy caterpillar : Aloa (Amsacta) moorei, A.albastriga
• Sunhemp hairy caterpillar : Utetheisa pulchella, A.astria
• Black Hairy Caterpillar : Amsacta lactinea</td></tr>
</table>

Aloa (*Amsacta*) *albistriga*

Spilosoma obliqua

Utetheisa pulchella

Noctuidae/ Agrotidae (Larges family)	• Largest family of Order Lepidoptera • Nocturnal, generally cryptic and dull coloured • Ocelli present • *Antenna* : Filiform • Maxillary palpi: Normally vestigial, labial palpi well developed • Proboscis present, rarely atrophied • Frenulum is present • Only 2 vannal vein in hindwings, 1st-2nd veins diverge near base of wings • *Fore Wings*: *Cryptic and dull coloured simulating with the surroundings*, M2 arises close to M3 than to M1. Cubitus appears four branched. In hind wings, Sc and R fuse for very short distance at the base of the discal cell.

	• **Larvae**: Usually 5 pairs of prolegs, but in some due to absence of prolegs on 3rd and 4th abdominal segments, a part of larval body forms a loop when moving, hence they are called *semi loopers*. • Most of them are highly polyphagous and nocturnal. Majority feed on foliage and some are stem borers, and they are called *army worms* or *cut worms* by their habit. • **Pupa**: Usually *pupate in earthen cell in soil* and the pupa is characterized by the presence of labial palpi and maxillae extending to the caudal margin of the wings. • ***Adults*** have a pair of well developed tympanal organs at the base of the abdomen. • Often adults are harmful, as fruit sucking moths, castor semilooper, *Achea janata* • Eggs are spherical, generally ribbed and reticulate **Examples** • Greasy/maize cut worm: *Agrotis ipsilon* • Spotted bollworm: *Earias insulana, E. vitella* • American bollworm : *Helicoverpa armigera* (also known as Gram pod borer/tomato fruit borer) • Citrus fruit sucking moth: *Eudocima* (*Otheris*) *fullonica*, *Eudocima* (*Otheris*) *materna, Eudocima* (*Otheris*)*ancilla* • Rice army worm: *Mythimna seperata* • Tobacco caterpillar: *Spodoptera litura* • Castor Semi Looper : *Acanthodelta* (*Achaaea) janata* • Cotton looper : *Anomis flava*

Mythimna seperata *Spodoptera litura*

Helicoverpa armigera *Anomis flava* *Agrotis* spp.

Acanthodelta janata *Erias vitella*

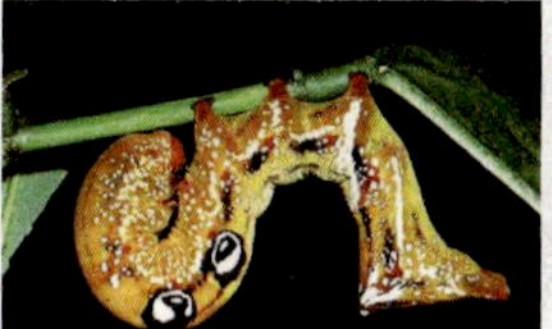

Eudocima materna *Eudocima fullonia*

Lymantriidae (tussock moths, gypsy moth)	• Medium sized, dull coloured, nocturnal moths with females of most species having only rudimentary wings Orgyia (= Notolophus) (*Notolophus* sp) proboscis is atrophied • Ocelli absent. • Antennae bipectinate in males and pectinate or plumose in females – sexual dimorphism • Wing venation resembles the Noctuidae. Sc and R fused to some extend and basal areole is larger in some species in hind wings • The caudal extremity of females is often provided with the large tuft of anal hairs which are deposited as a covering on egg masses • Caterpillars are densely hairy, often with thick compact dorsal tufts on certain segments • Osmeteria are frequently present on 6th and 7th abdominal segments. Some are provided with urticating's hairs which cause irritation • Pupation takes lace in a cocoon above ground and are characterized by specific evident setae • Apple gypsy moth : *Lymantria dispar* • Tussock caterpillar : *Olene mendosa* • Indian gypsy moth : *Lymantria obfuscate*

Euproctis fraterna *Lymantria obfuscata*

Olene mendosa

Lymantria marginata Lymantria todara Lymantria dispar

Tortricidae	• Codling moths(*Cydia pomonella*): apple pest of quarantine importance.

Cydia pomonella

Pterophoridae (Plume moths)	• They are small lightly built months • Forewings are elongate with two to four clefts or fissures. • Hindwings have three divisions • Legs are long, slender and armed with prominent tibial spurs. • Redgram plume moth, *Exelastis atomosa*.

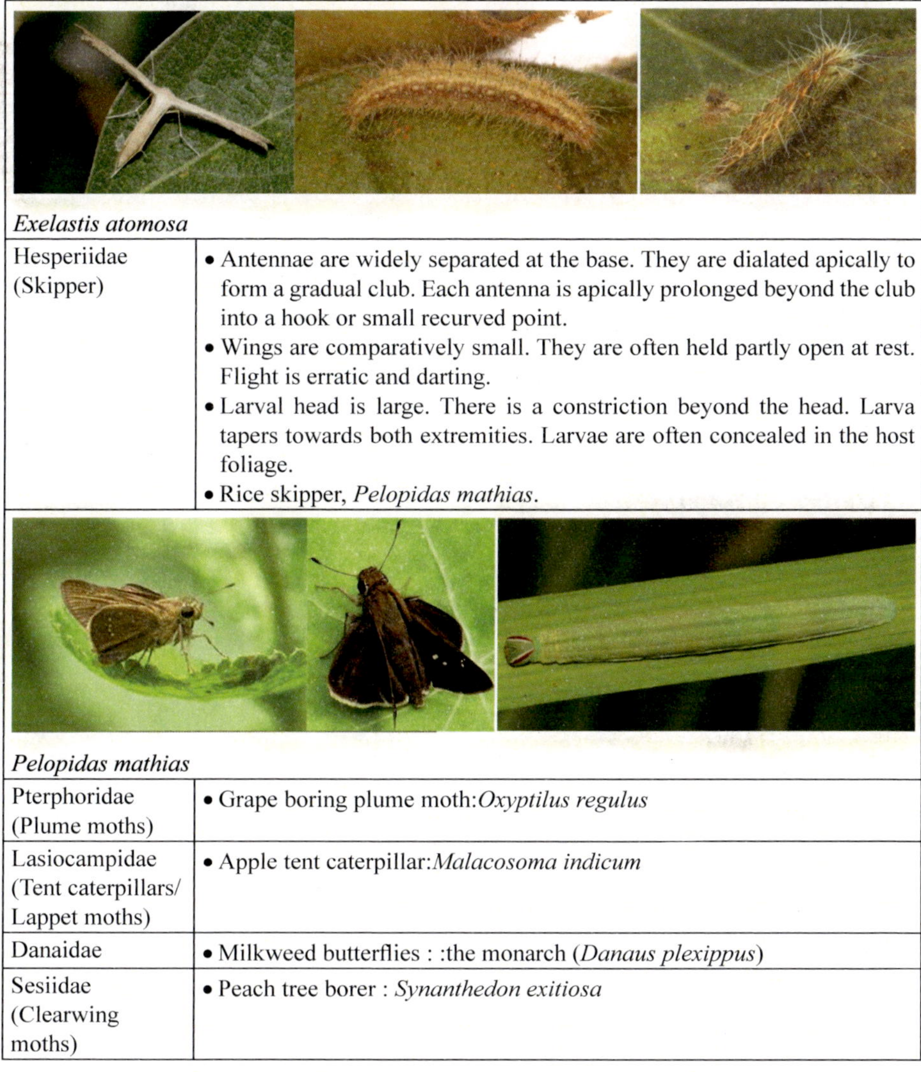

Exelastis atomosa	
Hesperiidae (Skipper)	• Antennae are widely separated at the base. They are dialated apically to form a gradual club. Each antenna is apically prolonged beyond the club into a hook or small recurved point. • Wings are comparatively small. They are often held partly open at rest. Flight is erratic and darting. • Larval head is large. There is a constriction beyond the head. Larva tapers towards both extremities. Larvae are often concealed in the host foliage. • Rice skipper, *Pelopidas mathias*.
Pelopidas mathias	
Pterphoridae (Plume moths)	• Grape boring plume moth:*Oxyptilus regulus*
Lasiocampidae (Tent caterpillars/ Lappet moths)	• Apple tent caterpillar:*Malacosoma indicum*
Danaidae	• Milkweed butterflies : :the monarch (*Danaus plexippus*)
Sesiidae (Clearwing moths)	• Peach tree borer : *Synanthedon exitiosa*

Diptera

(True Flies/Mosquitoes/Gnats/Midges)

The name Diptetra, derived from the Greek words "di" meaning two and "ptera" meaning wings, refers to the fact that true flies have only a single pair of wings.

- Head large freely moveable having small neck (cervix)
- Compound eyes large (Holoptic/Dichoptic); usually large in males
- Antenna 3-40 segmented, filiform, stylate or aristate (flies), plumose and pilose (mosquitoes).

- Mouthparts: Sponging type; piercing and sucking type in mosquitoes
- Mesothorax well developed. Prothorax and mesothorax being almost reduced to bands
- Tarsi 5-segmented and pads (pulvilli and an empodium usually present).
- Wings single pair membranous, hind wings modified into **halteres** (Gyroscopic).
- First abdominal segment is usually reduced. Number of abdominal segments varies. Musca 2-5 segments distinct. 6-10 segments form ovipositor.
- Metamorphosis complete.
- **Larvae** vermiform (maggots), apodus frequently with head reduced and retracted. Tracheal system most often Amphineustic.
- **Pupa** either free or enclosed in the hardened larval cuticle (**Puparium/ Coarctate**), adecticous, primitively obtect but in higher forms exarate.
- **Ptilinum (frontal sac)** indicated by the ptilinal/frontal suture is characteristic of Cyclorrhaphous diptera. It is used to force off the end of the puparium in order for the fly to emerge, and after this inflation at emergence, the ptilinum collapses back inside the head, marked thereafter only by the ptilinal suture (which defines the aperture through which it everts).
- Myiasis: Infestation of dipteran larvae in man and animals. *Musca domestica* and *Sarcophaga haemorrhoidalis* cause intestinal myiasis in humans.
- *Lucilia* causes sheep strike strike.
- Maggot therapy*: Lucilia, Phormia* used to clean wounds.
- Order Diptera is economically very important comprising of important crop pests, parasites, blood sucking mosquitoes acting as vectors of human diseases, predators etc. It is a large order, containing an estimated 2,40,000 species of mosquitoes, gnats, midges and others. They are an incredibly diverse group with a wide variety of fascinating life cycles and ecological adaptations; studying even one family of them is a life-time's work.
- The common names of true flies are written as two words, e.g., crane fly, robber fly, bee fly, moth fly, fruit fly. The common names of non-dipteran insects that have "fly" in their name are written as one word, e.g., butterfly, stonefly, dragonfly, scorpionfly, sawfly, caddisfly, whitefly.

The Diptera have traditionally been divided into **three Suborders**:

Characters	Nematocera (Thread horn)	Brachycera (Short horn)	Cyclorrhapha (Circular crack)
Antenna	Multisegmented, larger than head and thorax.	Shorter stylate, 3-segmented.	Shorter, 3-segmented
Arista	Absent	Present (Terminal)	Present (dorsal)
Palpi	Usually 4-5 segmented, pendulous.	1-2 segmented, porrect	1-segmented
Larva	Head exerted, biting mandibles horizontal.	Incomplete, retractile, biting mandibles, vertical	Mandibles vestigial.
Pupa	Free forms (weakly obtect)	Free/exarate except Stratiomyidae	Puparium/coarctate
Adult emergence	through a straight split in the thoracic region.	through a straight split in the thoracic region.	coarctate pupa has a circular line of weakness along which the pupal case splits during the emergence of adult. The split results due to the pressure applied by an eversible bladder **ptilinum** in the head.
Examples	Mosquitoes, Gall midges	Robber flies, House flies	Fruit flies, Agromyzids

Family	Features
Sub Order: Nematocera	
Psychididae (moth flies, sand flies)	• Blood sucking *Phlebotomus papatasi*, vector of Kala azar caused by flagellate protozoan *Leishmania donovani*.
Culicidae (Mosquitoes)	• They are delicate, fragile, slender insects • Females have piercing and sucking type of mouthparts with six stylets. • Legs are slender, delicate and long. • Males are short lived and feed on nectar or decaying fruits. • Females live long and are blood feeders • Sexual dimorphism (females antenna pilose/hairy; males antenna whorled/plumose). • Wings: 2nd and 3rd veins are forked, while the 3rd is unbranched and long. • Larva of mosquitoes are called **Wrigglers** having a well-developed head carrying dense tuft of hairs of feeding brushes over the mouth. Thorax is made up of a single fused segment and abdomen of 9 segments. • 4 tracheal gills present. • Pupae of mosquitoes are called as **Tumblers** having an enlarged cephalothorax which are pair of respiratory trumpets.

	• *Anopheles stephensi* and *A. culifacies*: Vector of human malaria and joint breaker virus. • *Culex fatigens:* Vector of Filiariasis caused by nematode *Wuchereria bancrofti.* • *Aedes aegypti:* Vector of Yellow fever virus and Chikun-gunya virus (Dengue virus). • *Mansonia:* Vector of Brazilian lethal illness.

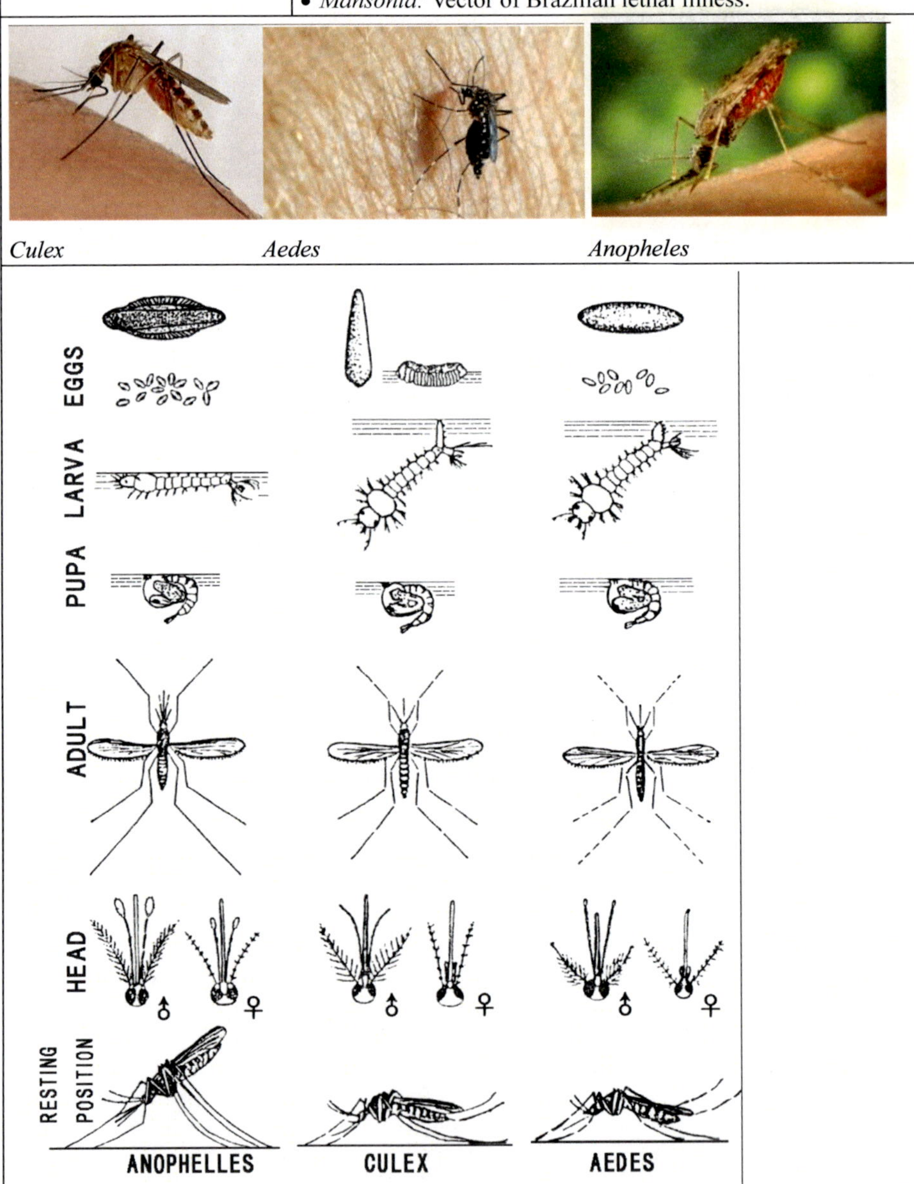

Culex *Aedes* *Anopheles*

Ceccidomyiidae (Gall midges, Gall flies)	• Minute delicate insects with long moniliform antenna with prominent whorls of hairs and legs. • Legs, long resembling mosquitoes, tibia without spurs • Ocelli present or wanting. • *Hairy wings*, unusual in the Order Diptera. Wing broad with • few longitudinal veins (3 to 5), most part unbranched. Cross veins are absent. • Larvae of most gall midges feed within plant tissue, creating abnormal plant growths called galls. • Larva : narrow at both ends and is characterized by the presence of a **sterna spatula or Breast** bone mid ventrally on the thorax. • Paedogenesis in *Miaster sp.* • Induce the formation of plant galls; others are scavengers, predators or parasites.
	Examples • Wheat Hessian fly: *Mayetiola destructor* • Rice gall midge: *Orseolia oryzae* • Mango gall midge: *Erosomyia indica* • Sesame gall midge: *Asphondylia sesami* • Sorghum earhead midge : *(Contarinia) sorghicola* • Midge on Chilli : *Asphondylia capsici*

Characters	**Anopheles**	**Aedes**	**Culex**
Eggs laying	Singly on or near the surface of water	Singly on or near the surface of water	Collectively in form of a compact mass or egg raft.
Egg shape	Boat shape with a float on either side	Ovoid having a series of small air chambers help in floating.	Fusiform
Larva feeding habit	Larva remains parallel to the surface of water with the help by palmate hairs.	Larva hangs head downwards inclined at an angle	Bottom feeders

Orseolia oryzae

Stenodiplosis sp.

Asphondylia sp.

Tipulidae	• Daddy long legs, crane flies.

Chironomidae	• *Chironomus sp.* : Larvae reddish (Blood worms) due the presence of Haemoglobin.

Simulidae	• Buffalo gnats
SUB ORDER: BRACHYCERA	
Asilidae	• They are elongate bristly flies. • Head is broad and hollowed out in between the compound eyes • Compound eyes are protuberant. A prominent tuft of hairs is found on the head forming the mouth-beard. The proboscis is thick and stout. • Legs are stout, hairy and suited for catching the prey.
	• Abdomen is tapering and has a pair of large claspers at the tip of the male and a horny ovipositor in female. • They are most active, non selective predators. • Robber flies
 *Robber flies*	

<table>
<tr><td>Tabanidae</td><td>• Body is stout
• Head is large. Eyes are large and often brilliantely coloured. In male eyes are holoptic (contiguous) and in female dichoptic (seperate). The third antennal segment is annulated. The proboscis is strong and pointing downwards.
• They are swift fliers.
• Male feeds on nectar. Female sucks blood from cattle and horses. They spread anthrax.
• Gad flies, Horse flies
• Tabanus speciosus: Vectors of Anthrax caused by Bacillus anthracis in domesticated cattle and man. Also vector of Tularaemia or ddr fly fever of man and vector of Trypanosomiasis, Loiasis.</td></tr>
<tr><td colspan="2">  
Tabanus sp.</td></tr>
<tr><td>Bombyliidae</td><td>• Bee flies</td></tr>
<tr><td colspan="2">SUB ORDER: CYCLORRHAPHA</td></tr>
<tr><td>Phoridae</td><td>• Humped backed flies</td></tr>
<tr><td>Syrphidae</td><td>• Hover flies or Syrphid flies
• They are birghtly coloured and brilliantly striped. A vein like thickening (spurious vein) is present in between the radius and median in the forewing.
• Abdomen has distinct black and yellow markings.
• Maggots prey on soft bodied insects especially aphids.
• Adults are excellent flies. They hover over flowers. They feed on pollen and nectar. They aid in pollination.</td></tr>
<tr><td colspan="2">  </td></tr>
</table>

Tephritidae (Fruit flies)	• Small to medium sized insects • Head large broad with small neck • Usually spots or bands on the wings. • Wing venation: Sub-costa bends apically forward at almost a right angle and then fades out without reaching the margins. • Female have sharp ovipositor and usually 3 segmented. • Middle tarsi with spurs • Adults are visitors of flowers, fruits and foliage • Larvae phytophagous, Amphipneustic (1st's last pair of spiracle functional) • Maggots can hop. • Cucurbit fruit fly: *Bactrocera (Dacus) cucurbitae* • Guava/Oriental fruit flies: *Bactrocera (Dacus) dorsalis* • Ber fruit fly: *Carpomyia vesuviana*

B. cucurbitae

B. diversa

B. latifrons

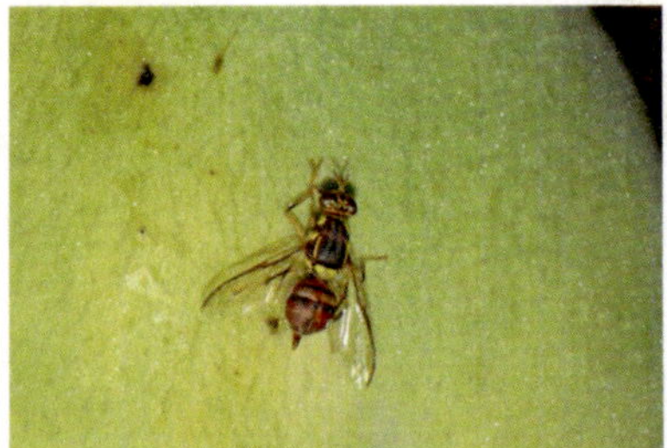

B. dorsalis

B. zonata

C. vesuviana

Agromyzidae (Leaf miner flies)	• Small to minute brackish or yellowish flies. • Larva leaf miners and characteristics in having spheroidal concretions of calcium carbonate in malphigian tubules. • Wings hyaline or pictured • Femora of legs bristled • Vibriosae are generally absent (a pair of stout bristles are on each side of the face just above the oral margin, longer than bristles on the vibrisal ridge) • Most of the larvae are phytophagous, mine into leaves producing characteristic blotches or mine into stems of young seedlings • Pea leaf miner: *Chromatomyia horticola* • Pea Stem Fly: *Ophiomyia phaseoli* • Red gram or Tur pod fly: *Melanagromyza obtusa* • American serpentine leaf miner : *Liriomyza trifolii*
Ophimyia phaseoli *Liriomyza trifolii* mines in leaves *Melanagromyza obtusa*	
Gasterophilidae (Bot flies)	• Horse Bot Flies: *Gasterophilus intestinalis*
Oestridae (Warble or Bot flies)	• Sheep bot fly: *Oestrus ovis*
Drosophilidae (Vinegar gnats, Pomace flies)	• Pomace fly : *Drosophila melanogaster*

Tachinidae (Tachinid flies)	• Ecto parasites of Lepidopteran, Coleopteran, Orthopteran insects (Biocontrol agents) • Small to medium, conspicuously bristly or hairy, active flies • Head is large and free. Arista on antennae often bare • Pteropleural bristles are present. Post scutellum is prominent • Wings are large , rarely mottled, Rs cell narrowed or closed apically • Arista is completely bare. • Abdomen is stout with several noticeable bristles. • Non specific endoparasites on the larvae and pupae of Orthoptera, Hemiptera,Lepidoptera and Coleoptera. • *Exorista sp*. on *Amsacta moorei* • *Strumiopsis inferens* on Rice stem borer, pink borer • *Sturmia bimaculata* – parasitoid on Spodopteera and other caterpillars • *Spogosia (Stomatomya) bezziana*- parasitoid on black heade caterpillar on coconut • *Exorista civiloides* – parasitoid on many caterpillars
   *Exorista Sturmiopsis* sp.	
Calliphoridae (Blow flies)	• *Chrysomyia bezziana* cause myiasis in man and cattle. Larva called as Screw worms.
Hippoboscidae	• Body is flat and leathery. • Legs are short, strong and useful for clinging to the host. • Wings are present or absent. • They are viviparous. They give birth to mature larvae which are glued to the hairs of the host. The young larva is retained in a special uterine pouch and nourished by special nutritive glands. Larva once laid never feeds. It pupates immediately. • They are blood sucking ectoparasites on cattle and dogs. *Hippobosca maculata* is associated with cattle and *H. capensis* is parasitic on dogs.
Muscidae (House flies)	• Small to medium dark flies resembling house flies • Antennal arista is plumose. • Mouthparts are sponging type. Labium is distally modifed into a pair of oval shaped fleshy lobes called **labella**. • Pretarsus consist of two claws and two adhesive pads. • First abdominal segment is yellow in colour. Terminal abdominal segments are telescopic forming a pseudo ovipositor. Abdomen is not bristly on basal part.

	• Fine erect hairs are present on the under surface of scutellum. Possess more than one sternopleural bristles • In the fore wing vein, Cu1 +1A is short and does not reach the wing margin • Larvae are cylindrical and truncated posteriorly • Maggots are scavengers. Adults carry certain disease causing microbes on its legs, body hairs and mouthparts. • Mostly phytophagous, some are scavengers and a few are parasitic • Diseases such as dysentery, cholera and yaws may be transmitted by pathogens carried on their feet and mouthparts. • Jowar shoot fly : *Atherigona varia soccata* • Rice seedling fly : *Atherigona oryzae*
 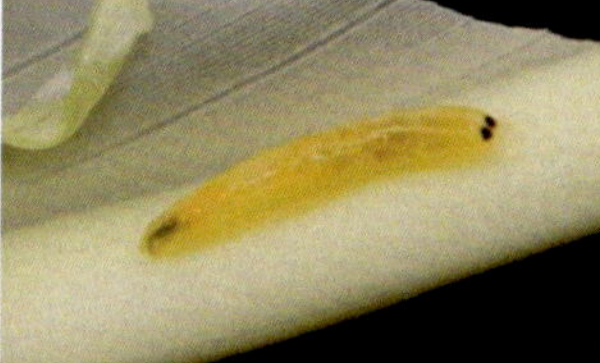  *Atherigona soccata*	

Difference

Hymenoptera

(Ants/Wasps/Bees/Sawflies/Horntails)

The name Hymenoptera is derived from the Greek word "hymen" meaning membrane and "ptera" meaning wings. It is also a reference to Hymeno, the Greek God of Marriage. The name is appropriate not only for the membranous nature of the wings, but also for the manner in which they are jointed together as one by the hamuli.

- The Hymenoptera is only order besides the Isoptera (termites) to have evolved complex social systems with division of labour.
- This is the most beneficial order in class Insecta comprising parasites, predators and bees involved in pollination and honey production
- Third largest order of Insects
- Head free from the thorax having neck (cervix); freely mobile
- Antenna variables. Geniculate in honeybees. Exhibit sexual dimorphism being longer in males.
- Compound eyes well developed. Ocelli are generally 3 in number. Acuteness of vision is a characteristic feature of the order.

- Biting and chewing mouthparts-except in bees (Chewing and lapping).
- Wings: two pairs of membranous wings often with veination greatly reduced; triangular stigma (Pterostigma) in front wings. Hind wings smaller than front wings, linked together by small hooks (hamuli). *Usually stigma is present in the fore wings* along the costal margin near apex.
- Legs modified for various purposes. In honeybees, Forelegs have Calcar for antenna cleaning. Tibia of hind legs-Pollen collecting type (Foragial type)
- Tarsi 5-segmented ending in aerolium and two claws. Parasitic forms-2-segmented trochanter (trochantellus)
- Narrow junction (Wasp waist ; Propodeum and Gaster/metasoma) between metathorax and 1st abdominal segment except in sawflies and horntails. The 2nd abdominal segment is long called *petiole or pedicel.*
- Larvae generally apodus
    - **Sawflies:** Eruciform (caterpillar-like); well-developed head capsule, chewing mouthparts; fleshy 8-pairs abdominal prolegs (Symphyta)
    - **Bees and wasps**: Grub-like, well developed head; chewing mouthparts, legless and eyeless (Apocrita)
    - **Parasitic wasps**: Body form highly reduced; lacking head, eyes or appendages (Apocrita)
- Pupae adecticous, exarate (rarely obtect) and a cocoon generally present except Honeybee, ants, chalcids.
- Digestive system: Crop modified as Honey stomach
- **Holometabola**
- Sex is determined by the fertilization of the eggs. Fertilized eggs develop into females and males are produced from unfertilized eggs. Males are haploid and females diploid.
- Hervivory is common among the primitive Hymenoptera (sub order Symphyta) in the gall wasps (Cynipidae), and in some of the ants and bees. Most other hymenoptera are predatory or parasitic.
- Most of the Hymenoptera have relatively unspecialized mandibulate mouthparts. An exception is found in the bees (as superfamily Apoidea) where the maxillae and labium are modified into a proboscis that works like a tongue to collect nectar from flowers. In these insects, the mandibles are used to gather or manipulate pollen and wax.

- Except for worker ants, most adult Hymenoptera have two pairs of wings. Front and hind wings are linked together by hooks (hamuli) along the leading edge of the hind wings that catch in a fold near the back of the front wings.
- In the Hymenoptera, females develop from fertilized eggs and males develop from unfertilized eggs. Since females control whether or not an egg is fertilized, they can regulate the sex ratio of their offspring.
- The fairyflies (family Mymaridae) are probably the world`s smallest insects. They parasitize the eggs of other insects.
- Some species of cuckoo wasps (family Chrysididae) invade the nests of wasp or bees, kill the larvae they find and deposit their own eggs on the stored provisions. This behavior is known as kleptoparasitism.
- Slave-maker ants raid the nests of other species to steal their pupae. When the stolen ants emerge as adults, they become workers in the slave-maker's colony.
- Aculeate Hymenoptera (certain wasps, bees and ants) are the only insects that can sting.
- Larvae of bees, ants and wasps do not form a complete digestive system until near the end of the pupal stage. Wastes accumulated by larvae are excreted just before the insect emerges as an adult.
- The females of some parasitic hymenoptera produce extremely large numbers of eggs. One Eucharitidae female was observed to lay 10,000 eggs in one hour.
- Fig wasps are the only insects that can pollinate fig trees. The wasp larvae, which developed in flower galls, become coated with fig pollen when they emerge as adults. They unwillingly cross-pollinate each flower they visit when laying eggs. The Smyrna fig is a commercial variety that does not produce any pollen. Its survival depends entirely upon *Blastophaga psenes*, a wasp that develops in wild caprifigs but cross–pollinate the Smyrna fig in a fortuitous case of mistaken identity

Economic importance

- Although some species are regarded as pests (e.g. sawflies, gall wasps, and some ants), most members of the Hymenoptera are extremely beneficial- either as natural enemies of insect pests (Parasitic wasps) or as pollinators of flowering plants (bees and wasps).

- The Hymenoptera are *exceedingly important insects* from man's point of view for three main reasons, Firstly they include the Bees who as everybody knows make Honey and Wax, secondly because a lot of the parasites and the Ants are important enemies of crop pests, ants consume huge numbers of lepidopteran caterpillars as well as other pests and were first deliberately used to protect plants from pests in China 4000 years ago when species of *Oecophylla* were encouraged to live in fruit trees because their presence was known to improve fruit yields. Hymenopteran parasites are regularly used in biological pest control these days and are among the first creatures screened by scientists when they are searching for a control mechanism for insect and other invertebrate pests. Thirdly and perhaps most importantly many Hymenoptera, though not the Ants are pollinators of most of our crop plants, both those we eat ourselves and those we use to feed our livestock. Because they work for free it is impossible to estimate their economic importance but it easily amounts to Billions of pounds every year. It is humbling to realise that if all the Hymenopterans were too suddenly dissappear from this earth among the numerous changes would be the collapse of human society
- Parasitism is a common way of life among a number of Hymenopterans

Classification

The Hymenoptera is divided into two Sub Orders:

Characters	Symphyta (Chalatogastra)	Apocrita (Clistogastra)
Members	Sawflies and horntails	Ants, bees, wasps, chalcids, Ichneumonids
Origin	Primitive	Advanced
Abdomen	Broadly jointed to thorax	Constriction between meta thorax and 1st abdominal segment
	No constriction between meta thorax and 1st abdominal segment.	Propodeum and Gaster present
Social	No	Highly social
Stemmata	Present	Absent
Legs	2 segmented trochanter (Trochantellus)	Normal
Feeding habit	Almost phytophagous	Mostly parasitic
Ovipositor	Normal	Well developed
Larva	Eruciform 3 pair of throracic legs + 6-8 pairs of abdominal prolegs without crochets. Pupate in silken cocoon.	Apodous/eucepahlous
Ovipositor	Normal- saw like suited for piercing the plant tissues	Well developed, suited for parasitism/ stinging
Behaviour sophistication	Less evolved	More evolved

Major Families

Family	Features
Symphyta (Chalatogastra)	
Tenthredinidae	• Only family with phytophagous insects, causing damage to various crops, due to caterpillar like active feeding larva • They are stout wasp like insects, often brightly colored without abdominal pedicel • Antenna : 3 or 6 or 8-11 segmented, filiform or setaceous. • Trochanter : 2 segmented, fore tibia with 2 apical spurs • Abdomen is broadly joined to the thorax • The ovipositor is saw toothed and suited for slicing the plant tissue. • Larvae is eruciform. It resembles a lepidopteran caterpiller. It has one pair of ocelli, papillae (reduced antenna) three pair of thoracic legs and 6-8 pairs of abdominal legs. Prolegs lack crochets. • Sawfly larvae also do not have velcro-like grippers at the ends of the prolegs. • Pupation occurs in an elongated oval silken cocoon or in earthen cell. • Larvae are external feeders on foliage. Larvae while feeding usually have posterior part of the body coiled over the edge of the leaf. • Parthenogenesis is very common. • Mustard saw fly: *Athalia lugens proxima*
 *Athalia lugens proxima*	
Apocrita (Clistogastra)	
Ichneumonidae	• Largest family of the Hymenoptera; parasitoids of other holometabolous insects (or spiders) • All parasitic on other insects • Large, slender, black yellow or reddish yellow insects • Antenna long > 16 segmented, filiform • Forewings has cross veins 2m-Cu. Two recurrent veins present. • In forewings, the costal cell is wanting and have two recurrent veins while braconids have one or none • Trochanter 2 segmented. Legs are provided with conspicuous tibial spurs and strong claws, tarsus 4 segmented • Abdomen long and slender, petiolate, petiole usually curved and expanded apically. • Propodeum is prolonged beyond the insertion of hind coxa and the gaster is generally 3 times as long as head and thorax.

<table>
<tr><td></td><td>• Ovipositor very long often longer than the body arising anterior to the tip of abdomen and exerted out permanently
• Solitary parasites
• Larvae endoparasitic/ectoparasitic and pupate in silken cocoon outside host.
• Hypermetamophosis. Caudal prolongation of tail is present, in 1st instar larvae
Examples
• Gambroides (Isotima) javensis: Parasitiod of Rice Yellow Stem borer and Sugarcane top borer.
• Xanthopimpla stemmator: Parasitoid of stem borer
• Eriborus trochanteratus is an exotic larval parasite of coconut black headed caterpillar.</td></tr>
<tr><td colspan="2">  
Xanthopimpla stemmator Isotima javensis Eriborus trochanteratus</td></tr>
<tr><td>Braconidae</td><td>• Very small stout bodied insects.
• 2m-Cu vein absent in the forewing. Only single recurrent vein present.
• The first sector of M+Rs present.
• Abdomen sessile or sub sessile or petiolate
• Costal cell is wanting in forewings with one recurrent vein
• Ovipositor well developed
• Petiole is neither curved nor expanded at the apex.
• Gaster is sessile or subsessile.</td></tr>
<tr><td></td><td>• Sternites of the gaster are partly membranous.
• Abdomen is as long as the head and thorax together
• Unlike Ichneumonids, many of these pupate in silken cocoons on the outside of the body of the host
• Poly embryony occurs in a few species of this family
• Parasitic commonly on lepidopteran larva
• Most braconids are primary parasitoids (both external and internal) on other insects, especially upon the larval stages of Coleoptera, Diptera, and Lepidoptera, but also some hemimetabolous insects like aphids. Most species kill their hosts, though some cause the hosts to become sterile & less active.
• Gregarious parasites</td></tr>
</table>

	Examples • Larvae endoparasitic (*Aphidius* sp.) ectoparasitic (*Apanteles* sp.) and pupate in silken cocoon. • *Apanteles flavipes*: Parasitiod of Jowar Stem borer (*Chiolo* sp.), PBW, Rice YSB. • *Cotesia (=Apanteles) glomerata* : Larval parasitoid of cabbage butterfly (*Pieris rapae*), cabbage looper (*Trichoplusia ni*), cabbage diamond back moth (*Plutella xylostella*) • *Apanteles africanus* : Larval parasitoid of tobacco caterpillar *Spodoptera litura* & other lepidoptran larvae • *Bracon brevicornis*: Parasitiod of Coconut Black Headed caterpillar (*Opisina arenosella serinopa*) • *Bracon hebetor* : Larval parasitoid of several species of pyralid moths, that attack stored products (Indian meal moth, *Plodia interpunctella*), nuts • Other parasitiods: *Chelonus* sp., *Aphidius* sp., *Stenobracon* sp.
   *Bracon gelechiae* *Apanteles Cotesia subandinus* *Bracon hebetor*    	
Cotesia flavipes *Chelonus blackburni* *Bracon brevicornis* *Aphidius colemani*	
Cynipidae (Gall wasps)	• Rounded elevation on scutellum and parasitic on puparia of flies • *Cothonaspis* sp.: Parasitiod of fruit flies puparia
Chalcididae	• They are small to medium sized insects. • Hind coxae are five to six times larger than forecoxae. • Possess 2 rows of short teeth on hind femur which is greatly swollen. • When at rest, wings are not folded longitudinally • Wing venation is reduced to a single anterior vein. • Prepectus is wanting • Ovipositor is straight and short. • *Brachymeria nephantidis*: Parasitiod of Coconut Black Headed caterpillar larva (*Opisina orenosella*)

Brachymeria sp.

Eulophidae	• Parasitiods of beetles, moths and other insects • Small insects, 1-3 mm parasitic on egg, larva and pupae of wide variety of insects • Tarsi 4-5 segmented • The axillae extend in front of anterior margin of scutellumand usually beyond tegulae • Antennae pectinate in males of many species • *Trichospilus pupivora*: Parasitiod of Coconut Black Headed caterpillar larva (*Opisina orenosella*) • *Tetrastichus ayyari*: Parasitiod of lepidopterous insects • *T. pyrillae*: Parasitiod of sugarcane pyrilla (*Pyrilla perpusilla*) • *Aphelinus mali*: Parasitiod of Wooly Apple Aphid (WAA) • *Encarsia (Prospaltella) perniciosi* belongs to family *Aphelinidae*: Parasitiod of Apple San Jose Scale
Trichogrammatidae	• Very minute insects 0.3 – 1 mm • Head short and somewhat concave behind • Abdomen constricted at base • **Tarsi 3-segmented** • Forewings broad with rows of microscopic hairs on them • Their forewings are typically somewhat stubby and paddle-shaped, with a long fringe of hinged setae around the outer margin to increase the surface area during the downstroke. Males of some species are wingless, and mate with their sisters inside the host egg in which they are born, dying without ever leaving the host egg. • They are not strong fliers and are generally moved through the air by the prevailing winds
	• Egg parasitoid of large number of insects. • *Trichogramma* has been the world's most widely used arthropod for augmentative biological control programs • *Trichogramma chilonis*: Parasitiod of YSB • *Oligosita nephotetticum*: Parasitiod of rice GLH

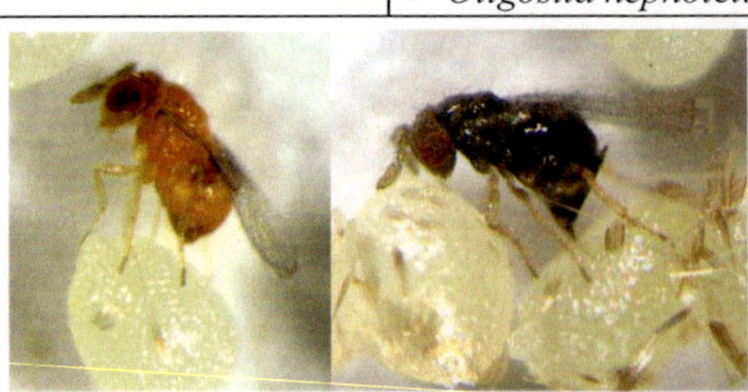

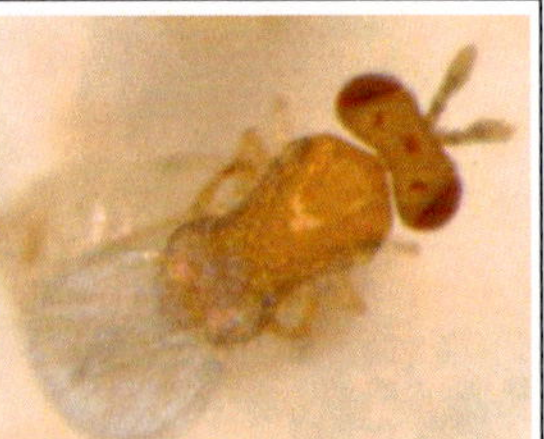

T. brasiliense T. japonicum T. embryophagum

Elasmidae	• Small brown/black elongated insects with enlarged, compressed hindlegs • *Elasmus* sp. • *E. albomaculatus*: Hyperparasitiod of *Apanteles malveolus*
Scelionidae	• Small black insects parasitic on egg, larvae and pupae of many insects • Phoresy: Transportation of insects by attaching itself to another • *Telenomus beneficiens:* Egg parasitoid of Rice YSB
Platygasteridae	• Minute black insects with 10 segmented antennae • Polyembrony noticed in some insects • Parasitiod of Aleyrodid and Ceccidomyids • *Platygaster oryzae*: Parasitiod of Rice Gall Midge
Formicidae (Ants)	• Social insects having queen, male and sterile caste (worker and soldier) • Antenna elbow (Geniculate) 1-2 segmented • Nodiform or scale like pedicel following the Propodeum • They are common widespread insects. • Mandibles are well developed. • Wings are present only in sexually mature forms. • Petiole may have one or two spines. • They are social insects with three castes viz., queen, males and workers. Workers are sterile females and they form the bulk of the colony. Exchange of food materials between adults and immature insects is common. After a mating flight queen alone finds a suitable nesting site. Wings break near the abscission suture near the base are nipped off by mandibles. Egg laying is started after divesting the wings. Usually the queen does not forage for food. During the initial phase of nest building it lives entirely on fat body reserves and products of wing muscle degeneration. • Many species have established symbiotic relationship with homopteran insects. • Mango Red Tree Ant: *Oecophyllas smaragdina*

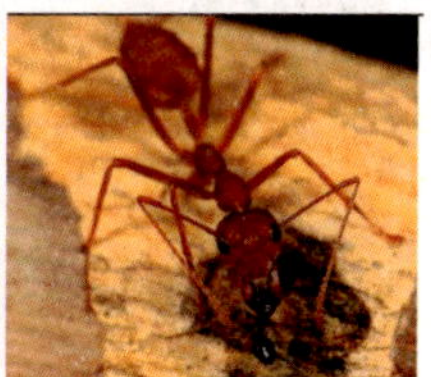

Oecophyllas smaragdina

Vespidae (Yellow jackets, Hornet, wasps)	• Yellow red insects with black marking • Social or solitary • Lateral extensions of the pronotum reach the point of insertion of wings and do not form rounded lobes. • Petiole long and slender • Forewings: very long 1st discoidal cell • Wing folded longitudinally at rest • Abdomen is conical • They construct nest with `wasp paper', a substance made from fragments of chewed wood mixed with saliva. • They are either solitary or social wasps. • They are generally predaceous on Lepidopteran caterpillars. Many paralysed caterpillars are stored in the cells of their nests. Eggs are suspended by a filament from the top of the nest and the cell is sealed • *Vespa sp., Polistes sp.* are predator of honey bees.

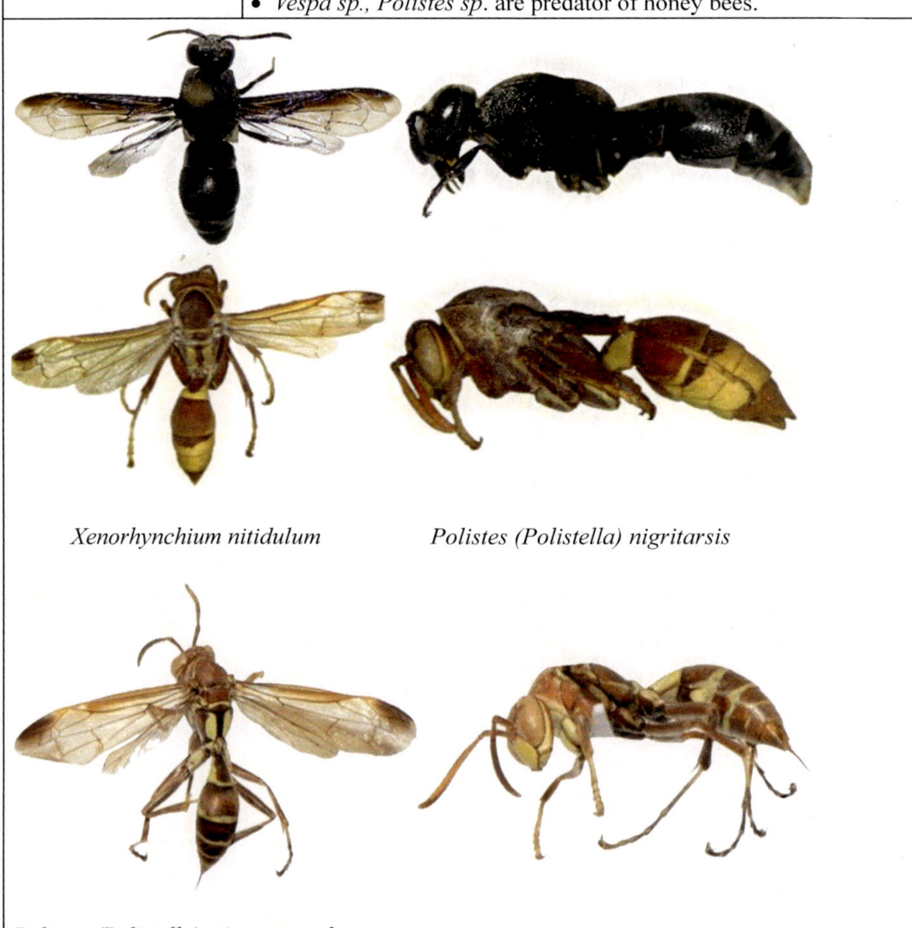

Xenorhynchium nitidulum

Polistes (Polistella) nigritarsis

Polistes (Polistella) stigma tumulus.

Apidae (Honeybees, Bumble bees)	Social bees play an important role in crop pollination • Bee caste: Queen, Drone (male) and sterile caste (workers) • Body hairs are branched or plumose • Three Sub-marginal cells in forewings • Antenna 13 segmented in male and 12 segmented in females • Hind legs: Pollen collecting (Foragial) type • Mouthparts chewing and lapping type • Honeybee devoid of spurs on hind tibia • Indian honey bee: *Apis cerana indica* (Domesticated) • Italian bee/Golden bee: *Apis mellifera* (Domesticated, introduced in India in 1962 by Dr. A.S. Atwal also known as father of Indian Beekeeping) • Little bee/dwarf bee: *Apis florea* (wild) • Giant bee/Rock bee: *Apis dorstata* (wild)

Apis dorsata

Apis cerana

Apis florea

Apis mellifera

Bumble bee

Xylocopidae (carpenter bees)	• They are large, robust bees. • Dorsum of the abdomen is bare. • Pollen baskets are absent in hindlegs. But brushes of hairs are present on hinglegs. • They build nests in dead logs and in live branches. They tunnel in all directions. They do not feed on wood. The tunnel is partitioned into several cells, separated by cemented wood chips. Pollen and nectar are placed in each cell together with one egg. The larvae hatch, feed, grow and pupate inside. • Adults are not aggressive and do not sting. They visit flowers and take nectar often by simply biting through the base of the flower instead of sipping from the top.

X. irridipennis *X. fesnestrata*

X. aeustuans *X.collaris*

Megachilidae (leaf cutter bee)	• They are solitary bees. • Mandibles are sharp and scissors like • Pollen gathering hairs (scopa) are present on the venter of the abdomen. • They cut circular or crescent shaped pieces of leaves of rose, redgram, guava etc. The cut pieces of leaves are used for preparing leaf lined cells. The provision for the brood consists of a mixture of pollen mixed with honey. • Rose leaf cutter bee: *Megachile anthracina* is a pest on rose, redgram and guava.

M. bicolor *M.disjuncta* Semi circular notches on leaves

Minor families	**Sawflies:** Larvae feed on foliage or burrow into plant tissues • **Diprionidae-** Conifer sawflies • **Cephidae-** Stem saw flies **Encyrtidae-** Mostly parasitiods of aphids and scale insects parasitic wasps. Larvae are parasitiods of other insects. **Predatory Wasps:** Adults provision nest sites with prey that they catch and paralyze by stinging. • **Sphecidae (Digger wasps)-** Prey on caterpillars and spiders. • **Pompilidae (spider wasps)-**Prey on spiders • **Tiphiidae (Tiphiid wasps)-** Prey on beetle larvae • **Scolidae (scolid wasps)-**Prey on beetle larvae • **Vespidae-** Yellow jackets, hornets, paper wasps **Solitary Bees:** Adults construct individual nests and provision them with plant materials (usually nectar or pollen). • Halictidae- **sweat bees** • Megachilidae- **leaf cutting bees** • Anthophoridae- **Carpenter bees**

Coleoptera

(Beetles/Weevils)

The name Coleoptera, derived from the Greek words "*koleos*" meaning and "*ptera*" meaning wings, refers to the modified front wings which serve as protective covers of the membranous hind wings. This is the largest order of insects (>3,60,000) comprising 1/4 or 25% of the known insect species..

- **Adults**
  - Size range from minute (0.25 mm) to largest (15 cm in cerambycids)
  - Head heavily sclerotized. Gula present (Beetles); absent in weevils.
  - Eyes usually well developed, except in Covernicolos and subterranean forms.
  - Mandibulate - Biting and chewing mouthparts (sometimes located at the tip of a beak or snout). Mandibles reach their maximum development in males of Stag beetles (Lucanidae).
  - Salivary glands absent in general in beetles.
  - Antenna varied (clavate, lamellate, moniliform, filiform, geniculate). Segments 1-27 or more (normal number 11).
  - Prothorax large and mobile, mesothorax much reduced and fused with metathorax and the tergum of these segments is divisible in to prescutum, scutum and scutellum.
  - A small part of the mesothorax known as scutellum remains exposed as a little triangle between the bases of elytra.

- Front wing (**elytra**) are hard and serve as covers for the hind wings; meet in a line down the middle of the back. Hind wings large, membranous, folded beneath the elytra or often reduced or wanting. Elytra firmly united and immovable (Carabids and weevils).
- Wing veination
    - Adephagid type- All veins and cross veins developed
    - Staphylid type- Cross veins absent
- Legs variable: Saltatorial (Alticids), natatorial (Dysticus sp., Grynius sp.), fossorial etc.
- Tarsi 2-5 segmented.
- Abdomen-Some of the proximal segments not clearly visible. Usually 10 segmented. First tergum is membranous and one or more of the sterna from the 1st to 3rd are aborted in many species. Terminal segments of female are retractile and tubular thus functioning as ovipositor (Family: Cerambycidae).
- Cerci and a distinct ovipositor are absent.
- Male genitalia typically a tripartite structure
- Stridulatory organs well developed in many beetles (Lucanids, Cerambycids)
- Pygidial glands secreting corrosive and pungent chemicals
- Malphigian tubules- Cryptonephridial type
- **Holometabola**

- **Larvae (Grubs)**
    - Head well-developed with ocelli and mandibulate mouthparts
    - Three pairs of thoracic legs; no abdominal prolegs
    - Body form:
        - Campodeiform- Slender, active crawlers
        - Scarabaeiform- Grub like, fleshy, c shaped body
        - Elateriform- Wireworms; elongate, cylindrical, with a hard exoskeleton and tiny legs
    - Pupa adecticous, exarate and are invested by a thin soft cuticle.

Economic Importance

Many beetles are regarded as major pests of agricultural plants and stored products. They attack all parts of living plants as well as processed fibres, grains and wood products. Scavengers and wood boring beetles are useful as decomposers and recyclers of organic nutrients. Predatory species, such as lady beetles are important biological control agents of aphids and scale insects.

Characteristics	Beetles	Weevils
Beak/Snout	Absent	Head prolonged in a beak or snout except Scolytidae
Gula	Present	Not developed and gular sutures are lacking or nearly always confluent
Labrum	Present	Usually wanting
Tarsi	5-segmented	Tarsi 4-segmented and in some 5-segmented
Thanatosis	Present	Absent
Example	Khapra beetle	Rice weevil

Coleoptera is divided into 3 suborders

Characters	Adephaga	Polyphaga	Achrostemmata
Hindcoxa	Immovably fixed to the metasternum completely dividing the 1st visible sternite	Not immovably fixed to meta sternum and not completely dividing the 1st abdominal sternite	Not immovably fixed to meta sternum and not completely dividing the 1st visible abdominal sternite
Wings	Usually show 2m-Cu cross veins		Distal part spirally coiled in repose
Prothorax	Noto pleural suture absent	Noto pleural suture absent	Noto pleural suture present
Larva	Mandibles devoid of molar areas	Mandibles having molar areas	Mandibles having molar areas
Legs	With a single tarsal segment ending in 2 (rarely 1) claw	With a tarsus and claws at least in 1st stage	With a tarsus and claws at least in 1st stage
Ovariole	Polytrophic	Acrotrophic	Acrotrophic
Examples	Tiger beetles, ground beetles	Pulse beetles, Weevils	Reticulated beetle

Family	**Features**
ADEPHAGA	
Cicindelidae	Tiger beetles-predators • Head is usually wider than prothorax. • Eyes are fairly larger and they have very keen vision. • Mandibles are sharply pointed, sickle shaped and acutely toothed for capturing the prey. • Legs are long and tarsi slender which enable to run fast. • Elytra have spots and stripes. • Larva excavates vertical pits for prey capture. • Both grubs and adults are active predators.
  *Cicindella* sp.	
Carabidae	• Ground beetles • Adults are often black in colour and some brightly spotted. • Some cannot fly because they have fused elytra and atrophied hindwings. • Legs are suited for running. • Larvae have caliper like mandibles, well developed legs and terminal cerci like structures called urogomphi. • They are nocturnal. Ground beetles are voracious predators both as adults and larvae. • They feed on soft bodied caterpillars and other insects. • Six spotted carabid : *Anthia sexguttata*
 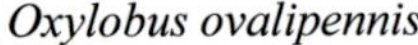 *Oxylobus ovalipennis*	 *Bradybaenus festivus*

  *Clivina assamensis*	 *Galerita orientalis Schmidt*
Dytiscidae	• True water beetles or predaceous diving beetles-large aquatic predators Body is long, oval, smooth and shiny. • Head, thorax and abdomen are compactly joined. • Antenna is filiform. • In some male beetles the foretarsi are provided with cup like suckers which are useful in clasping the mate. • Hindlegs are flattened, fringed with hairs and suited for swimming. • Air is stored beneath the elytra. • Adults and larvae are aquatic predators

Gyrinidae (Whirlinig beetles)	• They swim in erratic paths on water surface and exhibit gyrating motion. • Compound eyes are completely divided by the front margin of the head into an upper and lower half so that the beetle appear to have two pairs of compound eyes. The dorsal pair is suited for aerial vision and the ventral pair is for aquatic vision. • Forelegs are prehensile and long. • Middle legs and hindlegs are **natatorial**. • They are predators.

<table>
<tr><td>Hydrophilidae (Water scavenger beetles)</td><td>• They are black or dull coloured.
• Body is convex above and flattened below.
• Antenna is clubbed and kept beneath the prothorax.
• Maxillary palps are long and look like antennae.
• Legs are evenly placed in the anterior part of the body.
• Middle legs are flattened and suited for swimming.
• Metasternum is produced into a spine posteriorly.
• Air is stored beneath the elytra and over the undersurface of the body.
• Adults and larvae feed on decomposing vegetable matter.</td></tr>
<tr><td colspan="2"></td></tr>
<tr><td colspan="2">ACHROSTEMMATA</td></tr>
<tr><td>Cupesidae</td><td>• Reticulated beetles</td></tr>
<tr><td colspan="2">POLYPHAGA</td></tr>
<tr><td>Staphylinidae (Rove beetles)</td><td>• Elytra short. Veination M and Cu are not connected</td></tr>
<tr><td colspan="2">
Paederus sp</td></tr>
<tr><td>Galerucidae/ Chrysomelidae (Flea beetles, leaf beetles, Pumpkin beetles)</td><td>• Antennae are closely approximated, usually 11 segmented
• Third tarsomere is deeply bilobed
• Legs short. Hind femora enlarged for jumping in many forms. Tarsi 5 segmented, but appear to be 4 segmented since is small and concealed in the notch of the bilobed 3rd segment.
• Tarsal formula is 5-5-5. Usually tibial spurs absent.
• Abdomen short 5 visible sternites
• Larvae are root feeders.
• Adults bite holes on leaves.
• Red pumpkin beetles : Raphidopalpa foveicollis.
• Rice Hispa : Dicladispa armigera
• Grape flea beetle : Scelodonta strigicollis
• Sweet potato tortoise beetle : Metriona circumdata</td></tr>
</table>

<table>
<tr><td>
Aulacophora cincta</td><td>
Aulacophora foevicollis</td><td>
Aulacophora intermedia</td></tr>
<tr><td>
Scelodonta strigicollis</td><td>
Leptispa</td><td>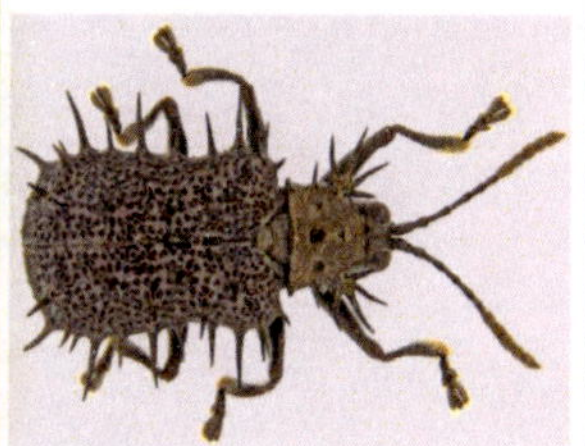
Dicladispa sp.</td></tr>
</table>

Melonothidae/ Scarabaedae (Scarab beetle, Cock chafers, June beetle, White grubs, lamellicorn beetles)	• Robust small to quite large • Head is small, broad and flat. • Labrum is well sclerotized • 3-7 segmented lamellate antenna • Tarsi usually 5 segmented and the foretibia often toothed or spinose with an apical spur • Mentum and ligula fused • Exposed Pygidium marks the 8th abdominal tergite • Adults attracted to light • Larva Scarbaeiform or C-shaped remain in soil • Both larva and adult are phytophagous and highly destructive • Common Indian dung beetle : *Heliocopris Bucephalus* • Sugarcane white grub: *Lepidiota mensuata* • The groundnut white grub: *Holotrichia consanguinea* • Coconut white grub: *Leucophilis coneophora* • *Leaf chaffer beetle : Anomala dimidiate*

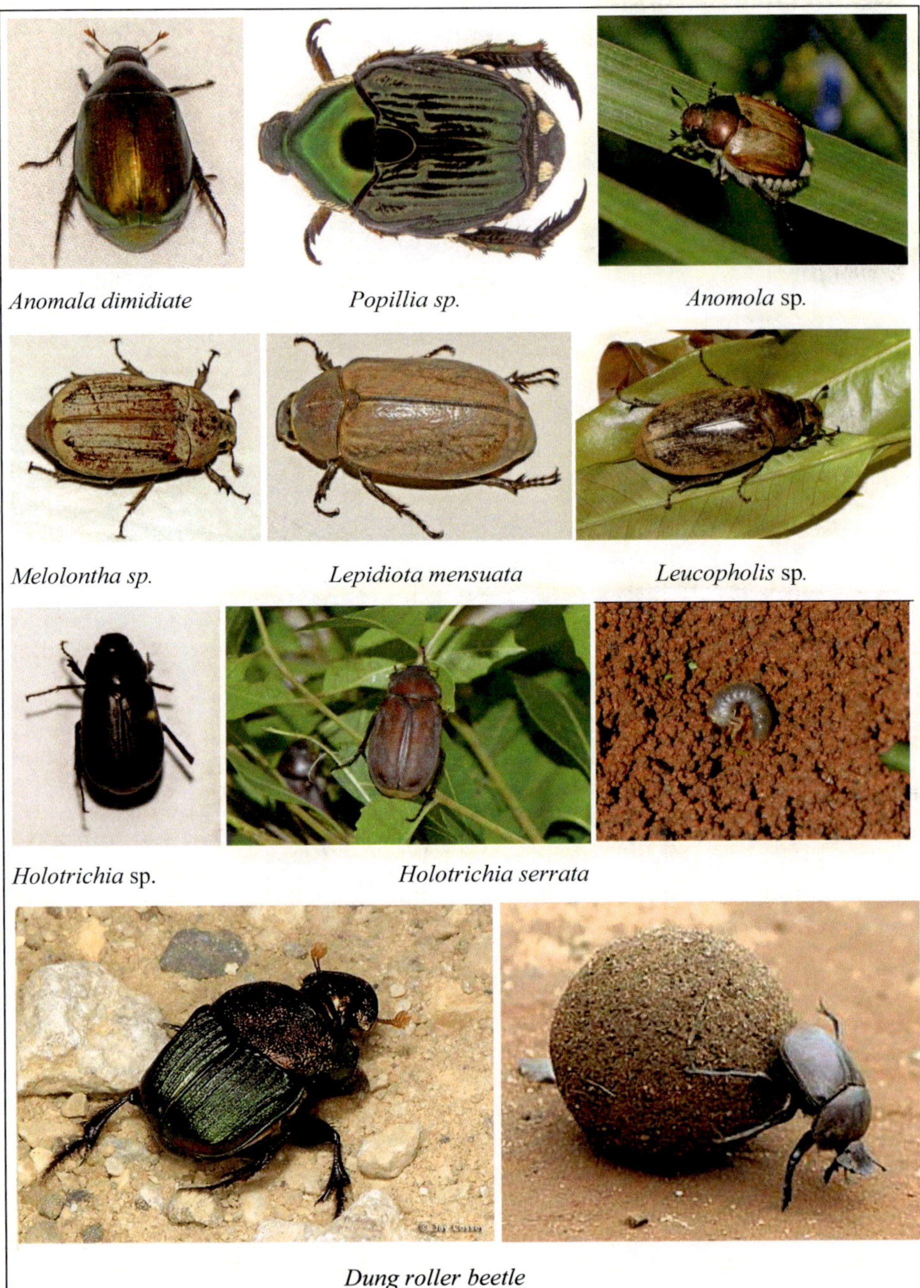

Anomala dimidiate *Popillia sp.* *Anomola* sp.

Melolontha sp. *Lepidiota mensuata* *Leucopholis* sp.

Holotrichia sp. *Holotrichia serrata*

Dung roller beetle

<table>
<tr><td>Dynastidae (Rhinoceros beetle, Unicorn beetle, Elephant beetle),</td><td>• Large to medium sized insects, smooth shiny, brilliantly coloured and sometimes sculptured
• Horns and tubercles are present on head and pronotum
• Mandibles are bent, expanded, leaf like and visible from above.
• Horns are usually present in male in the head and thorax.
• Sexual dimorphism
Examples
• Coconut rhinoceros beetle: Oryctes rhinoceros . Cephalic horns are found in both the sexes. In male the horn is longer and recurved. In female it is shorter and straight. Adults are injurious to coconut and grubs are found in dying palms and manure pits.</td></tr>
<tr><td colspan="2">  
Oryctes rhinoceros</td></tr>
<tr><td>Buperstidae (Metallic wood borers, jewel beetles)</td><td>• They are often elongate hard bodied insects.
• Body regions have a metaliic lusture
• Antenna is serrate.
• Larvae are called flat headed borers. Larval head is small and is entirely withdrawn into thorax. Prothorax is greatly expanded. Legs are absent. They tunnel beneath the bark or bore into stems or roots.
• Groundnut stem borer : Sphenoptera perotetti. The larva tunnels into the main root and kills the plants.
• Jute stem bupersitd: Agrilus acutus</td></tr>
<tr><td colspan="2">   </td></tr>
</table>

Lampyridae (Glow worms, Fireflies)	• The abdomen with photogenic organs in segments 6 and 7 in male and in segment 7 only in female, the light being stronger in latter • They show sexual dimorphism. • **Male** : head is concealed by the semicircular pronotum. • Eyes are well developed and contiguous. • Forewings are soft and flexible. They do not fully cover the abdomen. • Photogenic organ is found in sixth and seventh abdominal segments. • **Female** : Head is hidden by pronotum. • Eyes are very much reduced. • Wings are absent and is larviform. • Larvae are with sickle like mandibles. They are carnivorous and feed on snails. Extra intestinal digestion is common in larvae. • All life stages are luminous to varying degree. The luminescence is produced by the oxidation of a substance luciferin in the presence of an enzyme luciferase. The function of luminescence is to bring the sexes together. • *Luciola* sp., *Lamprophorus sp.*
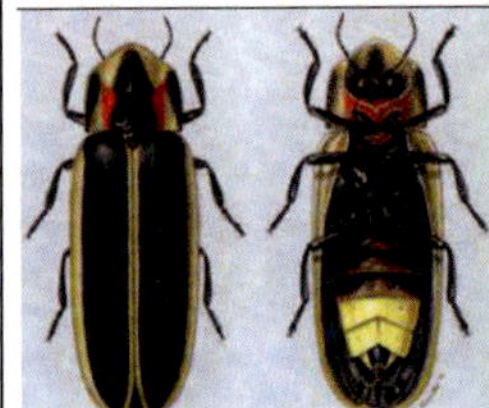	 
Dermestiidae	• Antenna clubbed, rarely serrate • 1-2 ocelli • Prothorax is not hood like • Adults hairy • Larva characteristics in having tuft of long hairs on their body • Khapra beetle: *Trogoderma granarium* (Stored grain pest, survive at moisture < 4 %)
 *Trogoderma granarium*	 

<table>
<tr><td>Anobiidae (Death watches)</td><td>• Body is oval shaped or cylindrical.
• Head is concealed by pronotum which is helmet like.
• Grub is fleshy with larger abdominal segments.
• Cigarette beetle: Lasioderma serricorne
• Drug store beetle: Stegobium panaceum</td></tr>
<tr><td colspan="2">
Lasioderma serriorne Stegobium panaceum</td></tr>
<tr><td>Bostrychidae</td><td>• Pronotum hood like
• Head hypognathus
• Antenna clubbed
• Elytra sloping and variously sculptured
• Coxae of forelegs are characteristics with reduced head and much enlarged thorax
• Larva with reduced head and much enlarged thorax
• Scobicia declivis: Lead cable borer or short circuit borer
• Lesser grain borer: Rhyzopertha dominica</td></tr>
<tr><td colspan="2">  
Rhyzopertha dominica</td></tr>
<tr><td>Coccinellidae</td><td>• They are hemispherical. The body is convex above and flat below.
• Their body appearance resembles a split pea.
• Head partly concealed by pronotum
• Elytra is strongly convex, brightly coloured, variously spotted and completely covered the abdomen..
• Tarsi 4-4-4, the deeply bilobed 2nd segment concealing the 3rd segment
• Grubs campodeiform elongated/flattened, usually bright coloured, spotted/banded and covered with minute tubercles/ spines.
• The last larval skin either cover the pupa or gets attached to the anal end of the pupa.
• Except the genus Epilachna, others are predators on aphids, scales, mites and whiteflies</td></tr>
</table>

	Examples • Rododolia beetle: *Rodolia cordinalis* predator of cottony cushion scale (*Icerya purchasi*) of cotton • *Menochilus sexmaculatus*: Predator of aphids, mealy bugs, scale, psyllids, whiteflies • Lady bird beetle (7 spotted beetle): *Coccinella septempunctata* feed on aphids • *Chilocorus suturalis*: Feed on aphids and coccids • *Chilocoris nigritus* and *Cryptolaemus montrouzieri*: Feed of coccids • Spotted leaf beetle (Hadda beetle): *Epilachna vigintioctopunctata* major pest of Brinjal

Epilachna vigintioctpunctata

Cheilomenus sexmaculata (Fab.)

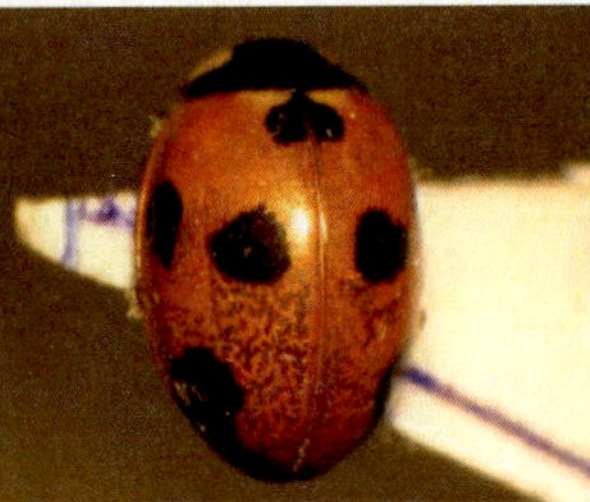

Coccinella septempunctata (Linn.)

Propylea dissecta (Mulsant)

Cheilomenus sexmaculata (Fab.)

Coccinella transversalis (Fab.)

Coccinella septempunctata (Linn.)

Chilocorus nigrita (Fab.)

Anegleis cardoni (Weise)

Unidentified

Tenebrionidae	• Body is flat or elongate • Winged or wingless. Elytra frequently fused • Antenna ending in abrupt club • Tarsi 5-5-4, heteromerous. Claws are simple • Larvae (**mealworms**) elongated and cylindrical and bear often two dorsal hooks on hind ends and ventral retractile process • Red flour beetle: *Tribolium castaneum*

Tribolium castaneum

Cerambycidae (long horned wood boring beetle, longicorn beetles)	• Phytophagous, highly destructive to trees • Compound eyes are notched • Antenna> body length. Antenna can be flexed backwards. It is surrounded at the base by compound eye. • Pronotum is with one to three laterally located spines • Tibial spurs well-developed • Tarsi 5-segmented, the 3rd being bilobed conceals the further segment. Tarsi 5-segmented but appear to be 4 segmented. • Tarsal formula 5-5-5 • Stridulatory organs (femora-alary or alary) • Grub-are called round headed borers apodus, elongate, cylindrical, whitish and bore in to hard stems, roots • Mango stem borer: *Batocera rufomaculata* • Grapevine stem girdler: *Sthenias grisator* • Coffee white borer: *Xylotrechus quadripes* • Longicorn beetles on cucurbits – *Apomecyna pertigera*

 *Batocera rufomaculata* *Sthenias grisator* *Xylotrechus quadripes* *Apomecyna* sp.	
Meloidae (Oil beetles, blister beetles)	• They are cylindrical, soft bodied beetles. • Head is connected to thorax by a distinct neck. • Legs are heteromerous with a tarsal formula of 5-5-4. • Claws show longitudinal splitting. • Forewings are soft and leathery. • Larval parasites, adult herbivores • Hypermetamorphosis • Beetles when disturbed, emit a fluid containing the oil principles "**Cantharidine**" which has irritating properties, through the openings in the apices of femur. • Adults feed on foliage and flowers • Blister beetle: *Mylabris pustulata*
Bruchidae/Laridae (Pulse beetle/Seed beetles/Dhora)	• Small short stout bodies covered by setae or scales • Elytra short, do not cover the tip of abdomen • Antenna serrate • Tarsi 5-segmented ending in claws hooked at base, the femora often swollen and dentate • tarsi formula 5-5-5, hind femur thickened and often toothed • Eggs are whitish, scale like and glued to the pods or seeds by a glutinous secretion. • Grubs feed exclusively on seed legumes. Pupation occurs within the seed. Adult emerges by cutting a circular exit hole. • Hypermetamorphosis • Pulse beetle: *Callosobruchus chinensis*, *C. maculatus*

Hispidae (wedge shaped or leaf mining beetle)	• Elytra with short spines and ridges • Larvae mostly leaf miners • Rice hispa: *Dicladispa armigera*
Apionidae	• Head is produced into a snout • Antenna is not elbowed. • Grubs are apodous. • Sweet potato weevil : *Cylas formicarius*. It attacks sweet potato both in fields and in storage.
 *Cylas formicarius*	
Curculionidae (Weevils, snout beetles)	• Minute to large sized insects. • Frons and vertex of the head are produced into snout. It is cylindrical and in some species larger than the beetle itself. • Mouthparts (Mandibles and maxillae) are present at the tip of the snout. It is useful to feed on internal tissues of the plant and provide a place for egg laying. Labrum absent. Palpi reduced & rigid. • In many species, it exhibits sexual dimorphism being better developed in females acting as a boring instrument for placing the eggs. • Antennae geniculate and clubbed arising about the middle of the snout • Palpi reduced and rigid, mouth parts small and arranged at the end of the snout • Wings : well developed, rudimentary or absent • Abdomen with 5 visible sternites • Legs: short or very long. Tarsi 5 segmented, 4th one often small. *Trochanter elongated.* • Wings well developed rudimentary or absent • Abdomen with 5 visible sternites • Adults and larvae are phytophagous and stored grain pests. • When disturbed they draw in their legs, antennae & fall to ground, remain motionless. • Grubs are usually apodous, curved with developed head (Eucephalous)

	Examples • Cotton stem weevil : *Pempherulus affinis* • Coconut red palm weevil: *Rhynchophorus ferrugineus*. • Rice weevil : *Sitophilus oryzae* • Sweet potato weevil : *Cylas formicarius* • Mango stone weevil : *Sternochetus mangiferae* • Banana rhizome weevil : *Cosmopolites sordidus* • *Banana stem weevil : Odioporus longicollis* • Myllocerus weevil : *Myllocerus undecimpustulatus* • *Bean weevil : Acanthoscelides obtectus* • Bean gall weevil : *Alcidodes* sp. • Coffee berry borer : *Hypothenemus hampei* • Alfalfa weevil/leucrne weevil : *Hypera postica*

Acanthoscelides obtectus *Alcidodes sp.*

Cosmopolites sordidus *Sitophilus oryzae* *Sternochetus mangiferae*

Tanymecus sp. *Rhynchophorus ferrugenius*

Pempherulus affinis *Odioporus longicollis* *Myllocerus discolor*

Hypothenemus hampei *Hypera postica*

Suggested Reading

Borror, Delong. J., Triple Horn, C. A. and Johnson, N.F. 1987. An introduction to the study of insects (VI Edition). Harcourt Brace College Publishers, New York, 875p.

BV David and Ramamurthy, V.V. 2016. Elements of Economic Entomology Namrutha Publications, 76/2, Sriramalu street Madanandapuram Chennai

Chapman RF. 1998. The Insects: Structure and Function. Cambridge Univ. Press, Cambridge.

David BV & Ananthkrishnan TN. 2004. General and Applied Entomology. Tata-McGraw Hill, New Delhi.

Duntson PA. 2004. The Insects: Structure, Function and Biodiversity. Kalyani Publ., New Delhi.

Evans JW. 2004. Outlines of Agricultural Entomology. Asiatic Publ., New Delhi.

Gullan, P. J. and Cranston, P. S. 2001. The insects- An outline of Entomology, II edition, Chapman & Hall, Madras, 491p.

Metcalf, C.K. and W.P. Flint. Destructive and Useful Insects : Their Habits and Control. Tata McGraw Hill Pub. Co., New Delhi

Nayar, K. K., Ananthakrishnan, T. N. and David, B. V. 1976. General and applied Entomology, Tata McGraw Hill Publishing Company Limited, New Delhi, 589p.

Richards OW & Davies RG. 1977. Imm's General Text Book of Entomology. 10th Ed. Chapman & Hall, London.

Romoser, W.S. 1988. The Science of Entomology, McMillan, New York, 449p.

Snodgrass, R. E. 2001. Principles of Insect Morphology. CBS Publishers and Distributors, New Delhi

Tembhare, D. B. 1997. Modern Entomology. Himalaya Publishing House, Mumbai, 623p.

Index

P

R

S

T